LEWIS, Richard S. The voyages of Apollo: the exploration of the moon.
Quadrangle, 1974. 308p il bibl 74-77942. 12.50. ISBN 0-8129-
0477-X. C.I.P.
There have been six landings on the moon by astronauts as part of the
Apollo program. The story of these explorations is well told in this
book by a veteran science reporter. Each trip is treated separately with
discussions of the voyage, the experiments and investigations carried
out on the moon, and the resulting scientific discoveries. Lewis suc-
ceeds quite well in pointing out how the experiments performed changed
our picture of the moon as new information was gained. We also get
some of the background as seen by a reporter who covered each mission. The work will be useful to introductory students and to others
who want a good, readable summary of the moon flights and their contribution to our understanding of the moon.

Richard S. Lewis, a native of Pittsburgh,
has had a long and varied journalistic
career as a newspaper reporter, science writer, and magazine editor.
Among his books are **Appointment on
the Moon,** a history of the space age
up to Apollo 11, to which **The Voyages
of Apollo** is a sequel, and **A Continent
For Science,** a history of antarctic exploration. His books in other fields are
The Nuclear Power Rebellion and **The
Other Child.**

THE VOYAGES OF APOLLO

QUADRANGLE

The Exploration of the Moon

RICHARD S. LEWIS

The New York Times Book Co.

All photographs and endpapers courtesy the National Aeronautics
and Space Administration.
Maps courtesy of NASA; adapted by Jack Jordan

Library of Congress Cataloging in Publication Data

Lewis, Richard S   1916-
    The voyages of Apollo.

    Includes bibliographical references.
    1. Project Apollo.   I.   Title.
TL789.8.U6A5453   1974   629.45'4'09   74-77942
ISBN 0-8129-0477-X

For Beatrice Lewis, in loving memory

Contents

~

(Illustrations follow page 212.)

Foreword

Long after the events that make today's headlines are forgotten, the Apollo voyages to the Moon will be remembered as the beginning of manned exploration of the Solar System. Seven expeditions were launched between mid-July 1969 and the end of December 1972. Six of them landed and carried out explorations, the first on another planet. I have undertaken to describe them not simply as a national adventure, as they were presented in most of the mass communications media, but as voyages of discovery in the context of a thousand years of exploration and intellectual growth.

As a result of Apollo, there have been fundamental changes in our outlook on two levels. From the Moon, we obtained a view of the Earth as a small, blue-white marble in a black sky. It provided visual confirmation of a new concept of the planet as a closed ecological system that man despoils at the peril of his survival. The image has enhanced the credibility of the whole environmental movement and the realization of the limits of resources.

As investigators continue to process data from the six Apollo landings, a new conception of the origin and evolution of the Moon has taken form. Tentative though it may be, it brings some order out of the confusion of pre-Apollo speculation. It provides constraints for models of the origin and evolution of the Moon and of the planets. This is certainly one of the great advances in the history of science.

Starting with Apollo 11, each expedition contributed data that, when processed in the laboratories, became the building blocks of the new Moon we see—or think we see—today. With each mission, new ideas arose and old ones were modified or discarded.

The exploration of the Moon has been a process of intellectual growth. That is essentially the contribution of the Apollo program. I have described this process as it developed, mission by mission, along with the models of the Moon that rose and set along the way.

We have seen the process before. In the fifteenth and sixteenth centuries, the European explorations of North America developed a conception of a New World very slowly. For generations, Newfoundland was thought to be a promontory of China. With each succeeding voyage, the shape of the New World became clearer.

So it has been on the Moon. With each Apollo landfall, the shape of that New World became more perceptible. If the New Moon that rises from Apollo explorations is still incomplete, it means that the exploratory effort is incomplete. In Apollo, we have only partially explored the Moon. It may be that our ideas about our New World are as far off the mark as those of sixteenth-century mariners and mapmakers were about the nature of their New World. But we have made a beginning. That is a story worth telling.

I cannot tell it, however, in the style of the great explorations of the past, for it does not happen that way.

In the past, even early in the present century, expeditions resolved themselves into contests between man and the elements.

In the lunar missions, man does not battle the elements—his technology does it for him. All his crises and narrow escapes are technical ones, often so complex that the details are left out in popular accounts. Our lunar explorers remain relatively comfortable in the microenvironments of their space suits no matter what is going on outside. When the temperature on the lunar surface rises above 200°F in the late lunar morning, our hero simply advances his cooling unit to medium cool.

In Apollo, the astronauts were once removed from the environment they were investigating. Instead of the elements, they had to contend with mechanical malfunctions and electronic "glitches." They survived because Apollo engineering represented the peak of mankind's technological development.

The space-age explorer is no longer the hardy adventurer, pitting brawn and pluck against a hostile environment. He is a highly trained, finely tuned technical expert, supported by a human pyramid of scientists, engineers, and technicians. It takes 5,000 people working around the clock to launch him to the Moon and more than 15,000 are involved in various phases of his journey. He is watched on television. Every breath he draws and every heartbeat are monitored. On the Moon, he is directed from Mission Control like a marionette by the electronic string of radio. Of course, he can exercise independent judgment if need be, but his schedule of activities is so tight and well rehearsed that he can barely draw an extra breath without falling behind the time line.

Under this regimen, the story of each mission is known in advance. It is prewritten in the flight plan that tells precisely what each astronaut is expected to do at a particular moment. It would be hard to find any aspect of human experience that is more regimented than manned lunar exploration.

But this is the way New Worlds are explored in the space age. There is no other way that works. The explorer himself is merely the extension of a vast, bureaucratic organism, which may be the National Aeronautics and Space Administration (NASA) or a similar agency in another country.

For this reason, accounts in the mass communications media have tended to concentrate on human interest aspects and logistical problems of the missions, rather than their detailed modus operandi and scientific results. The results always come later, months or years after the mission has ceased to be news. They have rarely been published in much detail. And because they often involve scientific concepts or relationships that must be explained for lay readers, they cannot be reduced, as more conventional human experience can, to a few paragraphs. It is not surprising, then, that the scientific results of Apollo leading to a new understanding of the processes of planetary development are confined for the most part to the scientific journals.

The lack of a broader and more intelligible report to a general readership amounts to an information gap. It may well account for the lack of public understanding and support of manned space exploration.

Is it reasonable to attempt to convey the detailed results of Apollo to a general readership? I think so.

The starting point is that the reader should be made aware that lunar exploration probes a new dimension in human experience. The language is often technical, abstruse, without analogy in the common knowledge, for in this dimension we seek to comprehend the processes of Creation.

The voyages of Apollo are a modern epic. But it is still so close to us that its heroic proportions are blurred. No doubt time will sharpen them. Then, perhaps, the exploration of the Moon will be seen in its real perspective, as an advance in the evolution of man.

R.S.L.
Evanston, Illinois
April 1974

THE VOYAGES OF APOLLO

1

The Exploration Imperative

For a thousand years the physical and intellectual expansion of people in the West has been marked by daring, often desperate voyages of discovery. These thrusts into uncharted seas and unknown lands reflect a process of growth inasmuch as they have always occurred about as soon as intellectual and technical development made them possible. Although motives have been variable through the centuries, the underlying drive to explore the environment and to expand the occupation of it is built into the genetic constitution of the human species. All history shows this. Modern man, *Homo sapiens*, evolved in one place and spread over the Earth, reaching the poles in the first half of the twentieth century and the Moon in the second half. The lunar landings in our time thus represent a point in a progression toward the exploration and eventual settlement not only of the Earth but of the larger environment of the Solar System.

Looking back a millennium, the roots of space exploration may be traced to medieval Iceland, the base for early European expansion into the Americas. The process did not begin there for Iceland was first settled by Irish monks seeking peace and solitude.

Tenth-century Iceland is a point in the progression from which it is possible to discern an acceleration of the exploration imperative with advancing social organization, technology, and competition. Evading the consequences of an Icelandic blood feud, Eric the Red sailed his Knarr* to Greenland, which was warmer a thousand years ago than it is now, and established a colony here. His son Leif explored the coast of

* A broad beamed vessel about 54 feet long with a big woolen sail.

North America in 1001, which the Norwegian trader Bjarni Jerjulfsson had seen when blown off course on a voyage to Greenland.

Five hundred years after Leif Ericson, when maritime technology provided larger ships and better navigation, Europeans were ready to cross the oceans, and they did so. The Age of Discovery opened. The voyages of Columbus in the fifteenth century, of Verrazzano, Frobisher, Drake, and Magellan in the sixteenth century, and of Cook in the eighteenth century paced the expansion of European commercial and military power into the Americas, Africa, India, Southeast Asia, Pacific Oceania, and Australia.

Five hundred years after Columbus, a similar pattern of exploration, energized initially by international competition, brought men to Antarctica, the last frontier on Earth, and landed them on the Moon, the first new frontier in the Solar System. In this context the lunar voyages of Apollo represent a continuation of Western man's intellectual and physical expansion from Iceland into the Solar System, consonant with his evolving technological and intellectual development.

Evolving out of the missile race with Russia and the intellectual stimulus of the International Geophysical Year (1957–1958), Project Apollo recapitulated the whole spectrum of motives for exploration in the last millennium. Military competition and national prestige, with their clear, economic implications, provided the fund-raising motives for Apollo, as military and economic considerations impelled the great voyages of Columbus, Magellan, Tasman, and Cook. Columbus sought a sea route to India that would give Spain a trade advantage over her Italian and Portuguese rivals. Magellan set out to discover a short route to the Moluccas. Elizabethan mariners searched for a northwest passage to Cathay. Captain Cook sought lands in the South Seas, including the fabled Southern Continent, thought to be as rich as the Americas.

In the eighteenth and nineteenth centuries new motives appeared in exploration mixed with the old. On his first voyage James Cook, then a lieutenant in the British Navy, sailed out of Plymouth, England in the summer of 1768 with instructions to perform a scientific mission—to observe the transit of Venus at Tahiti—after which he was to set sail in search of the Southern Continent. He found no trace of that mythical land, yet there it lay south of his track and hidden in mists and polar ice—impenetrable (to eighteenth-century ships) Antarctica.

As the less developed regions of the Earth were claimed for exploitation or colonization by the developed countries, territorial aggrandizement and exploitation of natives and resources gave way to the more refined considerations of national prestige, a powerful intangible among competitive nations, and intellectual curiosity, a new force in the world.

With the rise of natural science and philosophy came the great voyages of scientific discovery. They were sponsored or supported by the navies of the Great Powers, especially the British Navy, which found itself in need of missions after the Napoleonic Wars. Scientifically, the most rewarding of these voyages in the nineteenth century were those of H.M.S. *Beagle* (1831–1836), on which Charles Darwin developed his ideas about evolution and natural selection, and of H.M.S. *Challenger* (1872–1876), from which marine science and modern oceanography evolved.

In the same spirit, with an added dash of national rivalry, the polar regions were explored. After Robert E. Peary reached the North Pole in 1909, Robert Falcon Scott and Ronald Amundsen raced for the South Pole in 1911–1912. The contest lost its meaning once the Norwegians won it—just as the race to the Moon lost its fund-raising charm after the Eagle landed.

The age-old motif of national territorial aggrandizement and exploitation ended in the Antarctic. It was buried by the Antarctic Treaty of 1959, setting aside that region of 5.5 million square miles as an international preserve—a continent for science. Cynically, it can be said that the twelve nations signing the pact agreed to shelve their territorial claims for thirty years only after investigation during the International Geophysical Year showed that there was little of military or economic importance in Antarctica. But the treaty allows these claims to be reinstated in the future—and they may be if new technologies make the recovery of coal, oil, gas, and metals under the ice economically feasible. The Outer Space Treaty internationalizing the Moon and all other bodies of the Solar System beyond the Earth has proceeded from the same assumptions. In Antarctica and on the Moon the only apparent gain is new knowledge about the natural environment. Such spoils have never been considered worth fighting about.

Thus Apollo, designed to win the race to the Moon, fell back on its secondary motivation (from the viewpoint of Congress and the public) after Neil A. Armstrong and Edwin E. Aldrin landed the lunar module Eagle in the Mare Tranquillitatis on July 20, 1969. That was scientific exploration, a quest that would help man to understand the origin and evolution of the Earth and thus deal more reasonably with environmental problems.

Without the impetus of international competition, the intellectual motive was too unappealing to sustain a lunar adventure costing $2.5 billion a year. The expense became vulnerable to attack in a period of rising social and economic stress and political disaffection, aggravated by an unpopular war. These conditions were favorable for the resurgence of know-nothingism, and, in a burgeoning anti-intellectual climate, a counter-scientific revolution swept America, rejecting the values of science and ridiculing its goals. Social critics attacked the lunar adventure as irrelevant to the real needs of society. They wanted the money used to reduce the effects of poverty, ignorance, and economic disparity. Under these pressures and with public interest in the lunar adventure fading as its scientific complexity and technical proficiency increased, the last three of ten lunar landings originally planned were cut out of the program. So it was that the first manned exploration of the Moon began in 1969 and ended in 1972 with seven landing missions.* Six of the flights landed successfully. The scientific results of these missions have not resolved the origin and evolution of the Moon but have enabled us to interpret its dimensions.

The end of Apollo had always been predictable. The project had not been designed as a long-term one; it had been intended primarily to win the Moon race. The vehicle system was enormously costly. Each component—the three-stage Saturn 5 rocket, the Apollo spacecraft, and the lunar module—could be used only once. No rational program of long-term exploration could be carried on with such equipment. The planners and the architects of the national space program knew it. Immediately after Apollo 11 a presidential Space Task Group proposed a new system of reusable space vehicles, starting with a rocket airplane that would shuttle crews to and from an orbiting space station and including vehicles that could move men and equipment between the Earth's orbit and a lunar orbit and ferry them between the lunar surface and a space station in lunar orbit. This manned flight system could be developed to fly a crew to Mars. It projected a program well into the twenty-first century. The new program was initially much more costly than Apollo, but it would reduce the costs of exploring the Moon to a defensible level in the long term. In the harsh light of budget retrenchment, the long-range aspects of the program were shelved by the National Aeronautics and Space Administration and Congress.

Apollo had opened the door to manned exploration of the Solar System and proved that it was physically feasible. To many engineers,

---

* Two engineering test flights to the Moon preceded the first landing. Apollo 8, crewed by Air Force Colonel Frank Borman, Navy Captain James A. Lovell, Jr. and Air Force Major William A. Anders flew 10 orbits of the Moon at 69 to 71 miles altitude December 24, 1968. Apollo 10, crewed by Air Force Colonel Thomas P. Stafford and Navy Commanders John W. Young and Eugene Cernan circled the Moon 31 times at about 70 miles altitude May 21–24, 1969 for a test of the Lunar Module. Stafford and Cernan flew the LM to within 50,000 feet of the surface but did not land.

scientists, and astronauts who took part in Apollo, the significance of the lunar voyages was greater than their scientific and technical return. They marked a new era in the history of man—the beginning of human, physical presence in the Solar System beyond the Earth. If man could explore and ultimately colonize another planet, he could ensure his survival whatever happened to the Earth. He could reach for immortality, if he could divorce himself from total dependence on his natural habitat, or, to paraphrase the nineteenth-century Russian mathematics teacher Konstantin Tsiolkovsky, if he could escape from the cradle of his species.

The Steamboat Principle

All exploration in the last millennium has followed the Steamboat Principle, a concept first enunciated, supposedly, by Mark Twain: "When it's steamboat time, you steam." The idea implies the mystique of "readiness" in technological development. A society is impelled to move from one level of development to another when it is intellectually and technically ready to do so. This is analogous to individual human development. A child learns to read when he is ready—when he has acquired sufficient maturity and experience so that he can perceive, comprehend, and express the symbols of language.

Tenth-century Norwegians could not have reached or settled Greenland without a level of Norse shipbuilding technology and sailing technique that were probably the highest in Europe at that time. The building of larger and sturdier ships, the discovery of a method of divining latitude, and the invention of the clock, which made it possible to determine longitude, and the compass opened the world's oceans to navigation with a reasonable basis for survival. With the steam engine, nineteenth-century exploration and scientific enterprise moved very rapidly. The twentieth-century inventions of the ice-breaker ship, the tractor, and the airplane enabled men to occupy Antarctica in permanent settlements. Rocket development took them to the Moon.

Columbus, Magellan, Cook; Peary, Amundsen, Scott; Byrd; Armstrong and Aldrin—all investigated environments at an ascending level of technology. However, a gap of centuries may exist between the dawning of intellectual readiness and the technical capability of steaming.

The principles of space navigation have been known since the time of Newton and Kepler in the seventeenth century. Both gentlemen, after some emotional adjustment, perhaps, would have felt intellectually at home in the Apollo Mission Control Room at the Johnson Space Center,* Houston, where their ideas were applied, with some refinements, to send men to the Moon. The hydrogen–oxygen engine used to drive the upper stages of the Saturn 5 rocket was proposed 80 years ago by Tsiolkovsky. But the technology for building such a propulsion system was not at hand until the RL-10 hydrogen engine was developed for the Centaur rocket in 1963–1964. And the success of such an engine rested on a technique of electron-beam welding to build a tank that would hold supercold liquid hydrogen. This technique also was found to be effective a decade later in space-welding experiments aboard the Skylab space station in 1973. The tests showed that it is feasible to build large orbiting structures, such as a 50- to 100-man space station envisioned by the Space Task Group report of 1969, or a giant solar energy collector to beam power to a ground station.

The interdependence of science and technology is a recurrent theme in history. Without a bow-and-drill, flint and steel, a glass to focus sunshine or a match, one cannot build a fire. Without a computer, one cannot land men on the Moon where one wants them to land. And yet, without an analogous form of intellectual readiness for the experience, a landing on the Moon may be meaningless. Without experience acquired in exploring the Moon, further exploration into the Solar System may well be an exercise in frustration and confusion.

The Chemistry of Time

One of the great intellectual discoveries basic to any attempt at understanding the origin and evolution of the Moon—and all other bodies in the Solar System—was made two centuries ago. It was the discovery that the Earth was thousands of millions years old and not a mere 6000 years old as a strict interpretation of the Old Testament would have it. In shaping geophysics the conception of the great antiquity of the Earth was as important as the conviction that it was a sphere and that it moved in orbit around the sun.

In the Middle Ages learned men established the age of the Earth at five to six millennia. Medieval Jewish scholars dated the creation at 3760 B.C. The Jewish calendar still numbers its years from that date. In 1658 the Anglican Archbishop James Ussher pinpointed the creation at 4004 B.C.—at 8 P.M. on October 22, to be exact.[1] Greek Orthodox theologians set the creation at 5508 B.C.

The implication of these estimates is quick change—a sudden beginning and a sudden end to it all. The estimates were consistent with the

* Formerly, the Manned Spacecraft Center.

Old Testament account of the Creation in Genesis, which explained the origin of the Earth in terms of a 6-day project. Man appeared suddenly, fully developed, not unlike Athena springing spear and all out of the cranium of Father Jupiter. So it stood to reason that the features of the Earth, the mountains and valleys, rivers, lakes, and oceans also had been created by violent, sudden, catastrophic events, such as the Great Flood of Noah.

The idea that the Earth was 6000 years old or so and consequently had been shaped by catastrophic events persisted in the Western world until the eighteenth century. It began to erode away in the Age of Enlightenment. What an age that was—much more daring and more innovative than our own. It not only freed man politically from feudal institutions, but intellectually from the conceptual blinders that limited his ability to perceive and comprehend his natural environment.

In Scotland a naturalist named James Hutton (1726–1797) observed the effects of erosion and deposition upon the landscape. The mountains were worn away by wind and rain. Their tiny debris were carried by streams and rivers and deposited in the sea. This hardly perceptible shifting of mass on the planet, from higher to lower places, implied vast stretches of time. The rock series Hutton studied must have taken many more than six or seven millennia to form. In his book, *Theory of the Earth* (1785), Hutton noted: "There is not one step in all this progress that is not to be actually perceived." Such natural processes as erosion and deposition of rocks have been changing and forming the surface of the Earth at a constant rate throughout the entire history of the planet, he maintained. "What do we require? Nothing but time," he said.

Hutton's conclusions that the Earth must be very old indeed were supported by the Scottish-born, English naturalist Charles Lyell, the father of modern geology (1797–1875). Lyell saw what Hutton saw—the physical processes by which rocks were eroded, deposited in water, reconstructed and then, somehow, lifted up again to become new mountains. Enormous amounts of time as man reckoned it—by his heartbeat, by the length of the day, the lunar month, or the year—were required to account for the development of a landscape simply by the processes men could observe. The strictures of a brief life span for the Earth were broken. Scholars began to rationalize the widening rift between their observations and ecclesiastical authority by viewing the Biblical account as metaphorical. One day of the Creation might mean a millennium or an eon.

With the stretching out of time, a reasonable setting was created in which observed geological processes would fit. The work of Hutton and

Lyell established a conceptual basis for studying the evolution of a planet, for their work suggested clearly that the Earth had evolved and was continuing to evolve. The doctrine of change through observable, natural processes, such as the erosion of rocks by wind and water and the accumulation of sediments in basins and seas, became known academically as uniformitarianism. It displaced older, more dramatic ideas of rapid transformations, such as great sinkings, floods, and massive upheavals—a doctrine called catastrophism. True, floods, disastrous volcanic eruptions, great storms, and forest fires could be observed. These effects did not raise mountains, however, and they did not account for rocks with marine fossils on the ridges of high mountains. Other processes were at work that the eighteenth century did not perceive, and that the twentieth century perceives only dimly. The doctrine of change through time was the key to understanding the Solar System. Without this idea it would have been a waste of time and money from a scientific point of view to have explored the Moon, or any other planet. One simply could not have understood what one was looking at. It may be that mysteries on the Moon still eluding science require for their solution a new conceptual departure as radical as uniformitarianism was in the eighteenth century.

It was possible to measure time in the sedimentary rocks by counting layers. These could be differentiated by dark and light bands. The darker bands represented deposition of organic matter from dead leaves and bits of branches along with rock particles in the fall and winter. So the years could be counted and some idea gained about the number of years it took the rock to accumulate. Careful calculation of the time required for erosion and deposition to operate provided a record, which was marked by the presence of fossils. The oldest rocks with fossils in them appeared to be 500 million years old. They were found by Adam Sedgwick during the 1830s in northwestern Wales, representing a period he called the Cambrian (Cymry, the ancient name of Wales). In rocks older than the Cambrian series, no fossils were found. Fossil dating of the age of the Earth ended 500 million years B.P. (before the present) in Cambrian time. How long was Precambrian time?

Early in the eighteenth century an attempt to assess the age of the Earth was made by Edmund Halley (1656–1742), the English Astronomer Royal, the discoverer of one of the most conspicuous comets we see in the Solar System. Halley had a different system. By estimating the rate at which salt accumulated in the oceans from the erosion of surface rocks by rivers and their tributaries, Halley concluded that the Earth was 100 million to 1 billion years old.

Another approach in the mid-nineteenth century considered the rate at which the Earth had cooled from a molten birth. Physicists of this period were concerned with the sun's energy. Where did it come from? In 1854 Hermann Ludwig Ferdinand von Helmholz hypothesized that the sun gained energy by contracting its mass. If so, reasoned the British physicist William Thomson (Lord Kelvin), the Earth could not be more than 50 million years old.

He theorized that, in the process of emitting energy, the sun had contracted from a much larger mass, leaving the inner planets as molten blobs as it receded. Thus, the Earth began hot and has been cooling ever since. In this conception, Mars would be cooler, being "older," in the sense of having cooled sooner, and Venus would be much hotter than the Earth, being "younger." Kelvin refined his estimate of the age of the Earth toward the end of the nineteenth century to 20 to 40 million years. This conception was uncompromisingly at odds with that of the geologists, who thought the Earth must be hundreds of millions of years old in order to account for their observations.

The Radioactive Planet

In 1896 Henri Becquerel, a French physicist, noticed that radium emitted mysterious emanations that affected photographic film. He called them "radioactivity." Pierre and Marie Curie in 1903 found that radium was always hotter than its surroundings. It gave off energy. In a memorable lecture in 1904, Ernest Rutherford pointed out that radioactive minerals in the Earth were a source of heat. It was not necessary to imagine that the Earth had started as a molten blob of solar material and had been cooling ever since. The fact that the Earth became hotter at depth could be readily explained by the presence of radioactive elements in the crust. Instead of beginning as a molten mass, the planet could have accreted cold and then could have heated up from its radioactive burden. Kelvin's cooling calculations were thus rendered meaningless.

With the discovery of radioactivity, scientists not only found an explanation of the Earth's internal heat but they also acquired a clock that kept time on a cosmic scale. It could tell the age of the oldest and the youngest rocks. It could tell the age of the Earth itself.

Radiometric dating is based on the time it takes radioactive elements, such as uranium or thorium, to decay to "daughter" elements. Uranium and thorium decay to form lead; rubidium decays to strontium and potassium to argon or calcium. The rate of these transmutations is known from experiments. The most common uranium isotope, U-238, has a half-life of 4.5 billion years, give or take a few hundred million.

About half of it decays to lead (Pb) 206 in that time. The fissionable isotope, U-235, decays much faster. Its half-life is about 710 million years, by which time half of it is transmuted into lead (Pb) 207. The age of a rock can thus be determined by the ratio of "daughter" to "parent" radioactive elements. If the ratio of U-238 to Pb-206 is 50–50 in a rock, the rock presumably was formed 4.5 million years ago. The first objects dated at that age were meteorites. Some of them supposedly represent primordial condensation droplets or shards from the formation of the Solar System.

No rocks that old have been found on the Earth or the Moon at this writing, yet the two planets are believed to be as old as the meteorites. The younger rocks thus must be the descendants of older ones. They must represent cycles in which primeval rocks melted, cooled, recrystal- lized, and then underwent remelting and recrystallization. Radionuclide "daughters" that may have formed in the primordial or ancestral rocks were lost in the melting process, being more volatile than the "parents." The transmutation process started all over again when the rock recrys- tallized. The atomic clock was "reset" with each cycle of melting and recrystallization. In this way scientists can account for a 4.5 billion-year- old Earth on which rocks older than 3.7 billion years have not been found.

In potassium–argon dating, half of the isotope potassium (K) 40 decays to argon (Ar) 40 in 1.5 billion years. When a silicate melt cools, it starts to form a crystal, like an ice crystal. The argon present in the original material does not enter into the crystal, but the potassium does and becomes a part of the crystal. It then continues to decay at a known rate. The amount of argon-40 present can be assumed to be the result of potassium decay since the melt cooled.

The Sandbox

In any investigation of the Moon, radioactive dating is an essential tool, and without it the evolution of the Moon could hardly be fath- omed. The two indicators of time in terrestrial geology, fossils and sedimentary rock sequences, were missing on the lifeless, waterless satellite. Consequently, any investigation of the Moon was bound to be different from one on Earth. The classical basis of geology is a study of the major sedimentary systems of Europe and North America from the viewpoint of uniformity of processes and a 500 million-year time scale. On the Moon, however, it did not appear likely to scientists of the nineteenth and twentieth centuries that any analog of terrestrial sedimentary rock sequences had ever been formed. Possibly for this reason, specialists in Earth science paid scant attention to the Moon. It was generally considered to be in the province of astronomers. Nevertheless, a small group of lunar specialists agreed that the Moon probably had never had any large bodies of water nor much of an atmosphere. The processes of erosion, deposition, and sedimentation that require an atmosphere and a great deal of water in the form of rivers and oceans would be absent on the Moon. A new kind of geology—or selenology, as nineteenth-century geologists would have it—would have to be devised for the Moon. In many respects, the Moon was viewed as a primitive planet that had never evolved, with the homogeneous composition of a stony meteorite.

A widespread conception of the up-close appearance of the lunar surface was pictured by such artists as Chesley Bonestell. He painted stark and jagged landscapes, knife-edge peaks, and cliffs and scarps, unworn or abraded by erosion. For all of its beauty and imaginative power, this visualization turned out to be wrong. The gently rolling, tumbled contours of the lunar sandbox were a great surprise when the cameras were softlanded in Project Surveyor. One of the astonishing discoveries on the Moon has been the erosive effects of the solar wind, dust, and meteorites. Earth-based telescopes gave no hint of the gentle lunar landscape because they could not resolve lunar surface features that are smaller than about 1 kilometer. Resolution of detail was limited by the Earth's atmosphere. Small craters, down to several inches in diameter, that were revealed in the Ranger photos provided the first indication of surprises to come. When the Surveyor softlanders came down to reveal a surface resembling a freshly ploughed field in Indiana or Illinois, the new reality of the gentle Moon was confirmed.

Theories about the origin and composition of the Moon did not yield as quickly to the first contact, however. As Professor Harold C. Urey, a Nobel laureate in chemistry and dean of lunar scientists, remarked, proponents of conflicting theories could find support for their ideas in virtually every new discovery.

The Space Science Gang

During the early 1960s, as Project Apollo was being built, a group of researchers and theoreticians who began to think of themselves as space scientists was mobilized under the aegis of the National Aeronautics and Space Administtgation (NASA). Their spokesmen were highly optimistic about the prospect that an avalanche of new knowledge would cascade out of the heavens with the advent of probes and the manned space vehicles. The discovery of the Van Allen radiation zones, the solar wind, the shape and size of the magnetosphere, and the structure of the interplanetary medium gave promise of even more exciting things to come. "Within this decade, automated instruments will land on the moon and on some of the planets," said Lloyd V. Berkner and Hugh Odishaw in *Science in Space*, a book published in 1961 as a compendium of scientists' thinking about space research. Berkner, a radio engineer, had been vice-president of the Special International Committee of the International Geophysical Year (IGY). He became Chairman of the Space Science Board of the National Academy of Sciences. Odishaw, a physicist, was executive director of the Board. During the IGY, he had served as director of the World Data Center.

Science in Space expressed the new philosophy of space exploration: "Why is science concerned with these bodies? The planets of the Solar System are part of a whole—in their origins, their present states and in their futures. For the first time in history, man has it within his power soon to investigate these other bodies. And what he learns will be a significant addition to man's total knowledge."

In 1961 the revolution in geophysics arising out of discoveries during the IGY, especially in circumterrestrial space and in Antarctica, was at its height. Great advances were being made in the realization of the nature of the environment. The conception of man's natural environment was expanding from the Earth throughout the whole of the Solar System. Man was a solar being, not simply a terrestrial one. For did not the ultimate origin of all life go back to the sun, to the formation of the planets? Was not man a product of the chemistry of time that had shaped the Earth, the Moon, and the bodies of the Solar System? Could he not trace his descent from stars?

Science underwent a wave of popularization. New curriculum studies in earth science, physics, biology, and mathematics were financed by the National Science Foundation to improve science teaching in the secondary schools. In the mass media science writing became a popular feature. Public interest was directed to space exploration and the major focus was the Moon.

Because it was believed that the surface of the Moon had not "weathered" like that of the Earth, clues to the early history and development of the Earth might be found on the Moon, observed Berkner and Odishaw. "Even on the Moon, the remote possibility of some form of life cannot be ignored completely," they said.

The Key to the Solar System

There was a widespread expectation among theorists that the Moon might hold the key to the origin of the planets. Since the eighteenth century, it has been hypothesized that bodies of the Solar System condensed from a primordial cloud of dust and gas. This nebular hypothesis initially was proposed by the German philosopher Immanuel Kant (1724–1804) and the French mathematician Pierre Simon Laplace (1749–1827). Laplace visualized a flattened cloud of dust rotating about a center, throwing off, as it contracted, concentric rings of material. The rings coalesced into planets, about which smaller rings condensed into moons in some cases (Earth, Mars, Jupiter) or did not condense, as in the case of Saturn. Most of the cloud condensed to become the Sun. According to the hypothesis, the Sun should have most of the angular momentum (spin) of the entire solar system, but it does not. It has less than $\frac{1}{50}$ of the angular momentum of the planets moving around it. The discrepancy tended to defeat the whole, beautiful idea.

A variation of this hypothesis proposed by T. C. Chamberlin and F. R. Moulton of the University of Chicago in 1935 suggested that a passing star raised tides on the sun. Matter was ejected and condensed to form planetesimals. The larger ones swept up the smaller to form the planets. The Chamberlin–Moulton hypothesis accounted for the higher angular momentum of the planets by supposing the passing star had accelerated the ejected matter. Another variation was proposed by Sir James Jeans and Sir Harold Jeffreys, British astrophysicists. They supposed that a tidal filament was pulled out of the sun by a passing star. Jeffreys later imagined that a collision caused the filament to break loose. Modifications of these ideas characterize modern theory, which tends toward a more uniformitarian view that planets form about a star without the intervention of such catastrophic events as a collision or near collision of another star.

In other respects the nebular condensation idea suffers from a lack of credibility. A century ago, James Clerk Maxwell (1831–1879) argued that such a ring could not condense into planets, but would merely form small bodies, like those in the asteroid belt between Mars and Jupiter. Still another problem is why some bodies have different abundances of the elements than others. How could this be if condensation took place in a more or less homogeneous cloud?

It has been known for 70 years that the composition of meteorites differs from that of Earth rocks in a major way. The rocky meteorites, or chondrites, have more lighter, volatile elements, such as hydrogen, helium, nitrogen, and argon, than Earth rocks and a lower concentration of heavier, refractory elements, such as uranium or thorium.

From this arrangement, it can be supposed that meteorites have never been heated to the point where the volatile elements would boil off. That accounts for the lighter elements. But how can the depletion of the heavier elements be explained?

One explanation is that the chondrites were formed in a part of the solar nebula where the concentration of heavy elements was low. Since they had retained their volatile elements, the chrondrites must have been formed cold. So it seemed reasonable to believe that chrondrites coalesced at a much greater distance from the Sun than had the Earth. The heavy elements in which the meteorites were relatively deficient would have remained closer to the center of the solar nebula. The lighter, volatile elements would have been farther away.

This arrangement seemed to be confirmed by the order of the planets. The terrestrial or rocky planets—Mercury, Venus, Earth, the Moon, and Mars—were closer in; the gaseous ones—Jupiter, Saturn and Uranus— were farther out. Neptune seemed to be a ball of ice. Pluto was hard to account for unless it had spun off Neptune. In any case, this arrangement had a bearing on the origin and composition of the Moon. If the Moon turned out to be more like chrondritic meteorites than the Earth, it was likely to have retained its volatile elements and also to contain water. It was likely, too, to have condensed from chrondritic planetesimals farther out in the solar nebula than the Earth and thus could not have been formed near the Earth nor have been torn from it.

The chemical and mineral structure of the Moon in this way became an important clue to its origin. If the Moon was chrondritic and had formed cold, as Urey supposed, it must have formed elsewhere in the Solar System and have been captured by the Earth. If it was more like the Earth, it could have formed nearby or as a result of fission from the primeval planet.

The first rocks analyzed on the Moon might tell the story. On Earth, there were igneous and metamorphic rocks that had been cooked and recooked in the terrestrial furnace. During the cooking, the lighter elements, such as hydrogen, helium, argon, and oxygen, had boiled off, whereas the heavier or refractory ones, such as iron, nickel, gold, uranium, and thorium, had been concentrated into smaller rock masses, which we know as "ore bodies." In a molten Earth, the heavier elements, such as iron and the so-called iron-loving elements* (sidero-

* Nickel is an example.

philes), would sink toward the center, producing a nickel–iron core. Could these processes have occurred on the Moon? They would not have occurred if the Moon was chondritic and had formed cold. Lunar rocks would be high in volatiles, low in refractories, and would not show chemical differentiation from cooking that Earth rocks showed. The Moon would not have an iron core and consequently would lack a magnetic field like the Earth's, which is supposed to be produced by an iron core.

Even before the design of the great rocket that was to take explorers to the moon was completed, Urey and other scientists were talking about getting samples from the Moon for analysis. He and others proposed the development of equipment for the collection and return of lunar samples by automated space vehicles. Berkner and Odishaw suggested that the landing of men on the Moon would open it up to scientific exploration, as carried out on Earth. This prospect appealed to Berkner who had served as a radio man on the first antarctic expedition of Richard E. Byrd.[2]

In addition, Urey proposed that observations by a television camera on a lunar orbiting satellite would be useful. The satellite could be equipped also, with a mass spectrometer that would determine the presence and density of an atmosphere, a magnetometer that would tell whether the Moon had a magnetic field (which was important in determining whether it had an iron core), and a radiation detector that would tell whether an abundance of radioactive potassium, uranium, and thorium existed in the rocks. Urey proposed also a crashlanding probe with an ion chamber that would measure atmospheric density and radio the results until it hit. The probe would be instrumented to test the hardness of the surface, which some experts thought might consist of deep dust. The crashlander should be followed by a softlander equipped with a seismometer to detect and measure the intensity of moonquakes, a television camera to send back close-up views of the surface, a gravimeter to report the strength of gravity, and a radiation counter to detect any radioactive minerals at the landing site.[3]

Urey suggested that landings be made in the Mare Tranquillitatis, because of its black color and irregular shape which indicated a lava flow; the Mare Imbrium to the west, because of its gray color and obvious origin as a "collision" mare (that is, a basin sculpted by the fall of a huge meteorite or asteroid); the interior of a walled plain like Ptolemaeus; a mountain region covered by "ray" material from a great meteorite impact; the dark mountains to the west of the Copernicus; and the polar regions, where nuclides from beyond the Solar System and rare gases might be found. Samples of soils, rocks, and dust should be collected under strictly sterile conditions for later biological examination, Urey said.

Although Urey and others thought that automated collection would be a more likely possibility, they supported manned exploration. Urey said that the scientific qualifications of the first man to go to the Moon were very important. The investigator should have sound knowledge of several fields of science. He should be particularly a hard-rock geologist, with some acquaintance with meteorites. Urey assumed that without sedimentary rocks, the surface of the Moon would be more meteoritic than terrestrial.

These early suggestions were influential in shaping the scientific design of Apollo. In fact, with only a few changes, they became the blueprint for the landing missions of Apollo as well as for the Ranger, Surveyor, and Orbiter programs of lunar reconnaissance, which prepared the way for manned landings. One of the exceptions to Urey's recommendations was that the first man on the Moon be a hard-rock geologist. A geologist did go to the Moon. But he was not the first man to set foot on lunar soil. He was the last.

Notes

1. Asimov, Isaac, *The New Intelligent Man's Guide to the Sciences.* Basic Books, New York, 1965
2. Berkner, L. V., and Odishaw, H., *Science in Space.* McGraw-Hill, New York, 1971.
3. Ibid.

2

Models of the Moon

On the eve of Project Apollo, two major controversies existed concerning the Moon. One was its origin. The other was the manner in which the great mare basins, the craters, and the mountains had been formed.

Was the Moon a piece of the Earth's mantle, as its density suggested, ripped off during an early, plastic stage of planetary evolution? Was it a sibling of Earth, the junior partner of a double planet system that had condensed from a single cloud? Or had it coalesced elsewhere in the Solar System and been captured by the Earth as it swung by in an eccentric orbit around the Sun?

The question of origin was basic to the cosmogony of the Solar System. Thus, the investigation of the Moon was essential to the understanding of planetary evolution, which had become the most challenging question in natural history of the twentieth century.

If the Moon was a separate planet, had it evolved as the Earth had? Had it developed a crust, a mantle, and a core? Or was it an undifferentiated lump, like a meteorite? These questions seemed to rest on a determination of whether the mare basins and craters had been formed by impact of falling bodies or by volcanism.

From Earth, visual evidence supported both mechanisms. The dark areas of the maria, which had looked like seas to Galileo, looked like lava flows to twentieth-century geologists. This suggested either widespread volcanism or widespread impact heating. Some of the craters resembled terrestrial volcanoes. But extensive volcanism on the Moon did not jibe with Urey's early idea that the surface was meteoritic and hence low in the radioactive elements that would heat it up enough to produce volcanism. Urey had suggested that the Moon underwent a period of melting or partial melting early in its career and then became volcanically dead, but other observers insisted there was evidence of still active volcanism.

G. K. Gilbert's Moon

The classic exposition of the origin of the Moon's surface features was presented by Grove Karl Gilbert, senior geologist of the U.S. Geological Survey, in his address as retiring president of the American Philosophical Society on December 10, 1892.[1]

Like Hutton and Lyell, Gilbert attempted to reconstruct natural processes on the basis of evidence he could perceive. He was the first to deduce that successive levels of the Great Lakes were caused by the barrier of the receding glacier of the last Ice Age. The ice had cut off preglacial drainage routes.

In his address Gilbert said:

> The face which the Moon turns ever toward us is a territory as large as North America and, on the whole, it is perhaps better mapped. As its surveyor, even if armed with the most powerful of telescopes, is still practically several hundred miles away, his map does not represent the smallest features. . . . Upon his map are a score of great plains with dark floors, which he calls maria; there are a score of mountain chains; there are a few trough-like valleys remarkable for their straightness; there are many thousand circular valleys with raised rims which it is convenient . . . to call craters . . . there are thousands of bright streaks which are neither ridges nor hollows but mere bands of color; there are many hundred narrow linear depressions which he calls rills.

Gilbert observed the Moon on 18 nights in August, September, and October 1892 through the 26½-inch refractor telescope of the U.S. Naval Observatory. After reviewing theories of crater formation advanced up to this time, he concluded that only the impact hypothesis fitted his observations and deductions. He called it the meteoric theory.

He rejected the then prevailing notion that the craters represented a special type of volcanism peculiar to the Moon. Aside from the fact that the average size of the ten largest craters on the Moon is 24 times that of the ten largest ones on Earth, the structure of lunar and terrestrial craters appeared significantly different. "Thus, through the expression of every feature, the lunar crater emphatically denies kinship with the ordinary volcanoes of the Earth," said Gilbert.

He rejected also theories that the great craters on the Moon were formed by tidal effects of the Earth's attraction, by the melting of a snow and ice covered surface, or from the bursting of giant bubbles raised by expanding gases from boiling magma below the surface. Nor, he said, could they have been caused by projectiles hurled out of exploding volcanoes on Earth, a speculation that greatly underestimated the attraction of the Earth's gravitational field.

No—the only theory that made sense to Gilbert was the bombardment of the lunar surface by meteors. He reasoned as follows: "As the Moon either is without atmosphere or has one of extreme tenuity, the mechanical effect of this bombardment may be important, for the average velocity of the meteors is from 50 to 100 times as great as that with which the swiftest ball leaves the cannon. . . ." Even so, he admitted, it did seem incredible that the largest meteors known to have fallen on Earth produced scars the size of only the smallest observable lunar craters. Thus, he reasoned, it must be assumed that meteors of much greater size than the ones seen falling on the Earth in the nineteenth century bombarded both the Earth and Moon in the past.

If these rocks hit the Moon at a velocity of only 1.5 miles per second under the acceleration of lunar gravity, they would produce enough heat at impact (3500°F) to fuse the incoming body with the surface rocks. But if the incoming body was traveling at 45 miles per second (the average velocity of "shooting stars"), said Gilbert, "it is easy to understand that the heat developed by the sudden arrest of a fragment of rock traveling with such speed might serve not only to melt the fragment itself but also to liquify a considerable tract of the rock mass by which its motion was arrested."

Thus, the lavalike appearance of the crater floors might be explained by heat of impact. But one difficulty was not easy to explain—the circularity of the craters, which implied that the meteors fell vertically on the lunar surface. The difficulty was that the predominant angle of approach of fast-moving, meteoric bodies would be 45 degrees and at that angle of impact, the craters would be oval rather than circular.

Gilbert overcame this difficulty in an ingenious way. He theorized that the Moon coalesced from a ring of cosmic debris circling the Earth, like the rings around Saturn. A number of fragments or "moonlets" must have been left over from the early amalgamation. Gilbert surmised that the lunar craters were the scars produced by the collision of the leftover moonlets "which last surrendered their individuality." The velocities of these impacting moonlets would be so high that their courses in the vicinity of the Moon may be regarded as straight. "The introduction of the hypothesis of a Saturnian ring thus accomplishes much toward the reconciliation of the impact theory with the circular outline of the lunar craters," he said.

With great ingenuity and imagination, the dispute over the forces that had the main role in shaping the Moon's surface was continued into the twentieth century by the rival schools of volcanism and impact. Commenting on the volcanist point of view, Fred L. Whipple, director of the Smithsonian Astrophysical Observatory, noted "abundant evidence for volcanic activity."[2] He cited crater chains near the big Crater Copernicus, where volcanic sinks and domes are visible, and the Crater Wargentin, to the southwest, where it seemed that lava had been forced to the top of the crater walls, leaving them filled. Volcanoes like Vesuvius and Mauna Loa generally cause a mountain of material to rise above the central vent. Lunar volcanoes formed in this way are seen in the area of Copernicus at the edge of the Carpathian Plateau and in the region east of it. Although structures that seem to be volcanoes are relatively small in size, many volcanic effects such as cracking, melting, slumping, filling, and ridging are apparent. These exist not only in the maria but within filled craters and in flat regions between the mountains. Craters with dark halos caused by ejection of gases and slump craters seemingly caused by volcanic eruption appeared to Whipple to be abundant. Near the Crater Marius, in the western Oceanus Procellarum, serpentine ridges appear that certainly were formed by compression magma or lava flows. They showed volcanic domes 2 to 10 miles in diameter and 1000 to 1500 feet high.

On the other hand, Whipple thought that the larger craters on the Moon are all or mostly the result of meteorite impacts. In 1949 after the astrophysicist Ralph B. Baldwin amassed an array of evidence for meteoritic theory, Whipple felt that very few proponents of volcanic theory remained.[3]

Baldwin's Moon

From impact point of view, the Moon is a body sculpted primarily by the bombardment of planetesimals and meteorites. Here, indeed, is Catastrophism revisited. One of the few astrophysicists not in academic life—he was vice-president of the Oliver Machine Company of Grand Rapids, Michigan—Baldwin argued that most of the lunar surface features were formed early in the Moon's history by the impacts and subsequent explosions of myriads of meteorites, large and small.[4] "There is clear evidence of some sort of igneous action occurring as a secondary phenomenon" in the chain craters and lava flows, he admitted, as well as in craters having a central peak (or mountain) at their bottom. Nevertheless, despite these apparent igneous (volcanic) structures, the dominant process appears to be meteoritic impact. In his later book, *Measure of the Moon* (1963), Baldwin said he had found no reason to change his original views.

The formation of the moon's surface was a fundamental research problem for the Apollo missions. During the 1960s as Apollo began to take form, the rival theories of bombardment versus volcanism as the main force shaping the lunar surface were pursued with mounting vigor and excitement.

Baldwin argued that in the first billion years of the Moon's life, there were plenty of meteorites around that could have produced the landscape that is visible through field glasses. The Earth, too, was bombarded heavily during this early period, which may be the reason no terrestrial rocks older than 3.5 billion years have been found. Baldwin supposed that older rocks were pulverized by meteorite bombardment, which, because of the Earth's greater gravitational attraction, must have been even heavier than meteorite bombardment on the Moon. Subsequent erosion and melting wiped them out altogether.

What then became of Earth craters? Most of the earliest ones, which were probably the largest, vanished, erased by land-forming processes, slowly but inexorably. Pits were filled in. Crater walls eroded away. Only a few recent impacts are still visible, such as the mile-wide Barringer Crater near Winslow, Arizona, relatively new (only 10,000 years old), and the craters near Odessa, Texas, which may be hardly more than 11,000 years old. The Tunguska meteorite or comet that fell in Siberia in 1908 and the Karoonda meteorite fall in Central Australia in 1930 remind us that the impact process is still going on.

Meteorite bombardment, the impact theorists supposed, must be the way planets accreted in the first place. Urey estimated that millions of tons of mass are added invisibly to the Earth every year by the infall of micrometeorites—space dust.[5] What the lunar craters show, then, is the accretion process that has been preserved through time on a planet that is not as active as the Earth.

Beyond the recent meteoritic events, Baldwin suggests that many unusual, circular structures we see on Earth are actually ancient craters caused by meteorite impacts.

Cosmic Cannonballs

If the Earth underwent heavy bombardment in Cryptozoic (Precambrian) or Paleozoic time, would the evidence be virtually gone by now? Baldwin and others believe that some of it is still discernible on a close look. One of these old scars may be the Vredefort Dome in the Orange Free State, Africa. It is 75 miles across and certainly does appear to be the product of meteorite impact. So does the 3-mile Sierra Madre Dome near Stockton, Texas. Baldwin cites eleven structures in the American central plains that were probably meteorite craters. Among these are the Upheaval Dome in Utah; a deformed region around Kentland, Indiana; a deformed region centered at Desplaines, Illinois; the famous Serpent Mounds in southern Ohio; and Jeptha Knob in Kentucky. Canada, too, has a large share of ancient craters. Baldwin cites Deep Box Crater in Saskatchewan. It is 6¼ miles across, and 500 to 700 feet deep. Others have suggested that the southern part of Hudson's Bay is an impact structure.

The paucity of large meteorites in modern time, or in man's historic time, is cited in attacks on the impact theory. But the impact theorists argue that most of the big rocks orbiting near the Earth–Moon system were scooped up long ago by these planets. Only a few are left in the longer-lived orbits and these become subject to capture when their orbits are perturbed for some reason.

Baldwin pointed out that several small asteroids have been observed to pass perilously close to the Earth. One called Eros 433 comes within 14 million miles. Another, Albert 719, whizzes by between Earth and Mars in its journey around the Sun. Amor 1221 comes closer than either Eros or Albert. Another called Apollo passes within 1.8 million miles of the Earth and skims Venus by only 84,000 miles. Adonis has passed the Earth at 900,000 miles and Hermes has flashed by at 600,000 miles—a near miss, astronomically speaking. These bodies, except Eros, are about a mile in diameter. Eros is several times larger. Baldwin observed: "These cosmic cannonballs, each loaded, in effect, with high explosive, must be representative of thousands of such bodies." Each tiny asteroid would be capable of blasting a hole in the Earth's crust "entirely comparable to all save the very largest lunar craters."[6]

The great bombardments, however, ended billions of years ago, before men or vertebrate animals appeared on Earth. Time has nearly erased the scars on Earth but not on the Moon.

The Moon of Patrick Moore

The volcanic theory of lunar surface formation rests on a belief that the Moon began hot or became hot early in its history. One exponent of this view is Patrick Moore, a British astronomer. In a book published in 1963, he wrote: ". . . we begin with a Moon which has a solidifying crust lying over a hot, viscous magma."[7] Early in the Moon's history, it was closer to the Earth than it is now and the Earth's tidal effect on it was much greater. Lines of weakness in the lunar crust were produced by the Earth's tidal pull, after the tidal pull slowed the Moon's rotation to the point where one side always faced the Earth. At these weak areas, the hot magma under the crust pushed its way to the surface, raising a dome—as on the heated crust of an apple pie. Eventually melted rock and gas burst through. Then, with the relief of pressure below, the dome collapsed and fell back into the hot lava, to become remelted and eventually to cool into the dark, lavalike maria surface we see now. This was one hypothesis of the volcanists.

Moore reasoned that early lunar surface activity was the most violent and produced the largest craters. Thus, the largest craters are the oldest. On this point the volcanists and the impact people agree. The impacters contend the largest craters are the oldest because the largest meteorites struck first.

Dome collapses, said Moore, produced the great Mare Imbrium, which the impact school attributed to an asteroid the size of the Island of Cyprus. These collapses also formed the other major seas: Serenitatis, the great basin to the east of Imbrium; Nectaris in the southeast; Crisium in the east; and Humboldtianum in the northeast. Oceanus Procellarum, the huge, irregular basin in the west, might be the result of lava overflow from the collapse of the Imbrium Dome—or it may be a collapsed dome in itself.

Moore reasoned that as the great upheavals died down smaller craters appeared in the uplands and in the cooling floors of the maria. As time passed uplifts and collapses became less and less frequent. The inner magma cooled. Spiraling outward ever farther from the Earth, the Moon became less and less affected by the Earth's tidal effect and, at last, crater formation virtually ceased. Volcanists and impacters agreed that the processes that formed the lunar surface did wane in intensity with time.

As early as 1952 Moore predicted that the far side of the Moon would show fewer crater chains and large seas than the side facing Earth because the far side would have been less affected by tidal strain.[8] Seven years later the first photographs of the far side returned by the Russian Lunik III confirmed Moore's forecast. Later, photos by the American Lunar Orbiter cameras also confirmed that the far side is much less pocked with craters and does have fewer seas than the near side.

Other observers have pointed to the basaltic lava appearance of the maria as evidence of volcanism. In 1966 Jack Green, then a geophysicist with the Advanced Research Laboratory of the Douglas Aircraft Company, proposed that the maria were basalt, an igneous rock produced by volcanism on Earth. He suggested that reddish glows that

were observed coming out of the central mountain in the Crater Alphonsus, near the center of the near side, "fit the description of a volcanic eruption."[9]

Reddish or orange glows have been reported on the Moon for hundreds of years. Hot spots on the Moon show up in infrared telescopic
observations when the lunar disk is eclipsed. Investigators for the Air
Force Cambridge Research Laboratories charted them in observations
on December 19, 1964 and on April 13, 1968 during eclipses.[10] During the eclipse of December 19, 1964 also, the entire disk of the Moon
was scanned by investigators of the Boeing Scientific Research Laboratories, Seattle. They reported recording 563 "hot spots" on the face of
the Moon.[11]

The 1964 observations by the Boeing observers confirmed earlier
ones they made in March and September 1960. On March 13 that
year, observers using the 72-inch telescope of the Dominion Astrophysical Observatory (Canada) found that the Crater Tycho in the
south remained warmer than the regions around it when the lunar surface cooled during an eclipse. There was evidence, also, that the Craters
Aristarchus in the northwest and Copernicus in the midwest cooled
more slowly than their surroundings. The hot spots were noted in most
of the big ray craters—the ones from which rays of lighter material
seem to emanate and extend vast distances.

To the volcanist school, these observations were strong evidence for
continuing volcanism on the Moon. But whether volcanism was the
dominant force shaping the lunar landscape or simply an auxiliary one
remained to be proved. Certainly the existence of the "hot spots" and
the basaltic appearance of the maria suggested strongly that volcanic
activity had played a role in forming the lunar surface. Also, whether
the heat of volcanism had been produced by the decay of radioactive
elements in the Moon, or whether it had been generated by the energy
of impacting meteorites remained a controversial issue. The resolution
of the controversy not only had a bearing on the origin of the Moon and
its evolution but on the mechanism of planet formation and development.

Harold Urey's Moon

On the eve of the launch of Apollo 11, the first attempt to land men
on the Moon, Urey summarized his position of volcanic versus impact
theories of lunar formation as follows[12]: "Today, it is generally agreed
that the great craters of the Moon were produced by collisions of meteor-

ite or asteroidal objects which arrived on the Moon at some time during its long history." There was also a history of volcanism on the Moon. Some people, he added, "prefer to believe that this consisted of great lava flows which filled the maria regions."

In Urey's view, impact destroyed the entire early crust of the Moon —and the same process must also have wiped out the earliest igneous and sedimentary rock formations on Earth. This, he reiterated, explained why we do not find rocks on Earth older than 3.7 billion years. All the Earth rocks formed in the first billion years of the planet's history (4.5 billion years) were pulverized by the initial bombardment, melted by Earth's volcanism, and recrystallized as younger rocks. Thus, the bombardment that destroyed the early surface of the Earth also destroyed the early surface of the Moon. The question was: Did the melting and recrystallization of rocks that took place on the Earth also happen on the Moon? By 1969 there was strong evidence that such a process must have occurred on the Moon, for a chemical analyzer that was landed in the Mare Tranquillitatis aboard Surveyor 5 on the morning of September 10, 1967 sent back data indicating that the dark rock was similar but not identical to terrestrial basalt, a product of volcanism.

Urey speculated further that it was possible that the bombardment record exhibited by the Moon shows us how the Earth originated, too. Although volcanic processes may be present on the Moon, he conceded, they could not have had a major role in shaping it. The Moon simply was never hot enough. In 1968 tracking data from the Lunar Orbiters indicated that large concentrations of mass (the mascons) existed under the surface of circular maria, suggesting that lunar rock structure is as strong as steel. Urey reasoned that if the interior had been hot, the rocks would have been too soft and too plastic to have supported the mascons so near the surface and to have maintained the Moon's irregular shape. He concluded that the Moon could not have been formed in a molten state, nor could it have been very hot early in its history.

If the Moon was never very hot, it should have at least as much volatile material, such as hydrogen, helium, and oxygen, as meteorites. It should have a lot of water, Urey supposed. That was a possibility of high interest to Project Apollo. If there was water in the Moon and if it could be tapped, it would be an invaluable resource for lunar settlement. It would provide a source of breathing oxygen as well as oxygen and hydrogen for rocket fuel and for fuel cell batteries supplying electricity.

If the Moon had water, Urey did not rule out the possibility that sedimentary rocks, like the sandstones and shales of Earth, might have formed early in the Moon's existence. "It is my belief that water was briefly present as a surface material on the Moon early in its history."[13]

But there was no evidence of valleys of terrestrial type, where water washes mountains down to the sea in about 20 million years. It would be interesting to see if Apollo 11 brings back any hydrated rocks, he said. That would prove the existence of water.

As the argument about impact versus volcanism proceeded, it became increasingly obvious that it was inextricably bound up with the question of the Moon's origin. Both questions clearly generated a third: Where did the bodies that bombarded the Moon so catastrophically early in its history come from? The origin of these projectiles was a piece of the whole question of Solar System evolution. So was the question of whether the Moon was a blob of plastic matter spun off the Earth in the beginning, whether it was a separate body that condensed elsewhere in the Solar System and was captured by the Earth, or whether it accreted from a leftover ring of dust and gas and assorted debris around the Earth.

Urey believed the Moon was captured. It represented a class of objects that may have formed the nucleus for the accumulation of planets. Otherwise, it was difficult to see how the planets would accumulate by themselves.[14] "The asteroids are presently disintegrating due to collisions and I see no reason why asteroidal sized objects in the past would not have disintegrated in the same way," Urey said. "However, if we had a solar nebula consisting of asteroidal objects and we tossed in a few objects of lunar size, the planets would accumulate without any trouble at all."* From this view, the Moon was a protoplanet that never grew.

The question of how accumulation took place worried physicists considerably. The mechanism was not clear. How did the smaller objects— and the lunar-sized ones—accumulate? Were there gravitational instabilities in the solar nebula that might account for such condensations? Gerald P. Kuiper, who headed the Lunar and Planetary Laboratory at the University of Arizona, suggested that protoplanets of considerable mass were accumulated directly from the nebula by gravitational instabilities. He supported the meteor-impact theory of lunar surface formation, with Urey.

There was an element of wishful thinking in the capture theory. If the Moon had been formed independently of the Earth, Urey said, it would

---

* The solar nebula is generally conceived of as a primordial ringlike cloud of dust and gas from which the planets and lesser bodies of the Solar System condensed some 4.5 to 4.6 billion years ago.

be much more interesting to investigate than if it *were* merely a spun-off fragment of the Earth. "Stepping on the Moon would have the same interest as stepping on Mars, or the asteroids, or Venus."

Gerard Kuiper's Moon

In 1954, when he was Professor of Astronomy at the University of Chicago and director of the Yerkes and McDonald Observatories, Kuiper put forth a theory not only covering the questions of the origin and evolution of the Moon but also dealing with the sources of bodies that had bombarded it. Kuiper reasoned that the Moon must have melted as a result of radioactive heating, a process that he regarded as uniform.[15]

If one assumed that the Solar System was 5 billion years old, it could be assumed also that early in its history the production of heat by uranium and potassium-40 was 10 times greater than now (inasmuch as radioactive decay had depleted the original amount of these elements in the solar nebula). In the early Solar System, then, spheres of condensed matter would have been heated to the melting point in less than a billion years, he estimated. All spheres larger than 100 kilometers in diameter would have melted at least close to the center. Larger ones would have melted entirely, or nearly so, leaving, perhaps, an outer shell a few kilometers thick of unmelted but probably plastic material.

As a result of convection and cooling, the radioactive elements would have been brought near the surface layers—as on Earth. No further general melting would have occurred (because near the surface the heat would be radiated more readily into space).

Telescopic observation of the Moon for decades tended to support the idea that it had undergone a thermal cycle. For there in the great maria basins, which Kuiper supposed had been scooped out by planetoid impacts, was the dark material that resembled lava.

In view of the Moon's thermal cycle and because lunar gravity was "not negligible," Kuiper thought it could be assumed that the Moon had "at least a small iron core." If that was so, the density of the lunar surface material must be less than 3.34 (times that of water), which was the average density of the whole Moon, inasmuch as the density of the center would be considerably higher.

To Kuiper, it seemed plausible that the Earth and Moon originated as a double planet system within a common envelope and not from separate clouds of matter. After several hundred million years during which heat accumulated from radioactive decay, the Moon began to melt outward from its center. Only the crust remained solid for a few kilome-

ters, but it did not escape thermal alteration. From a premelting stage of
loose rubble it became hardened by metamorphism. The maria, Kuiper
suggested, were formed at the stage of peak melting that coincided with
the massive bombardment that formed the basins. The impacts of meteorites, planetoids, or asteroids pierced the crust and allowed the lava
to well out upon the newly formed basins.

Telescopic observation also showed that the maria were not created
all in one catastrophic event—but over time, although the time was
limited. During the maria-forming epoch, the undersurface was molten
and the melt poured out each time a hole was punched in the crust, just
as boiling juice flows from a punctured apple pie. It appeared to Kuiper
through the telescope that Mare Serenitatis had flowed into the Imbrium
Basin and thus was younger than Mare Imbrium, and Mare Tranquillitatis flowed toward Serenitatis and must be younger than Serenitatis.

Although the impact hypothesis provided an acceptable explanation
to Kuiper for most of the larger basins and craters, volcanism also
played a major part in shaping the surface. Kuiper divided the craters
he had studied into two classes: those formed before the melting process had encompassed the Moon and those formed after the melt congealed. Looking at the bright full Moon, Kuiper noted that some huge
craters, such as Clavius, were nearly invisible and others displayed only
ghostly outlines. He supposed that these were premelting features,
largely filled in by the melting process. The rayed craters, such as
Copernicus, Kepler, and Tycho, were formed after the melting epoch.
Their raylike splashes of debris, which had been punched out by impacts, could be seen overlying the dark maria lavas in the form of
lighter streaks.

Kuiper observed that the postmelting craters appeared a bright white
at full Moon, whereas the premelting ones were barely visible. This
demonstrated, he said, that the crust had undergone metamorphism
during the melting epoch.

The distribution of the "white" craters showed that the melting was a
general phenomenon and was not limited to the surface layers of the
maria. "In other words, the heat causing the melting of the maria was
not primarily derived from the kinetic energy of impacts but was largely
preexisting at the time of the impacts and was due to a general heating
of the Moon by its own radioactivity," he said. Thus the maria were the
products of central heating.

Ammunition Belt

Next Kuiper addressed himself to the question of the origin the projectiles that bombarded the Moon. He listed four possibilities: (1) they formed by condensation in the inner part of the Earth–Moon cloud; (2) they formed inside the proto-Earth but outside the original Moon's orbit, implying that this part of the proto-Earth was too tenuous to become gravitationally stable and formed a second, outer satellite around the Earth; (3) they formed between the Earth and Mars; and (4) they formed in the asteroid belt (between Mars and Jupiter), coming to the Moon as meteorites or asteroids.

Kuiper ruled out the asteroid belt as the source of the bombardment. Because of the increasing rate of pulverization in the asteroid ring, the present rate of bombardment from this source is greater than the rate during the early period of the planetary system. But the Moon shows an enormous impact rate early in its history and a very low rate later, so that the early bombardment must have come from another source. The most likely source, said Kuiper, was Process 2 in which moonlets condensed from the proto-Earth outside the orbit of the Moon and eventually fell into it in the final chapter of lunar accretion. Kuiper's Process 2 harked back to Gilbert's Saturnian-ring conception. Some of the Process 2 moonlets were of considerable size. The one that created Mare Imbrium was 150 kilometers in diameter, Kuiper calculated.

Kuiper's 1954 theory suggested a visualization of an early Earth, a nearby early Moon, and a ring of sedimentary debris beyond in orbit around the double planet system. As tidal effects caused the Moon to spiral away from the Earth, the Moon eventually reached the orbit of the sediment ring, where it was battered catastrophically by chunks of congealed matter falling on it at relativity low velocity. This battering created the great basins. Kuiper has the Moon passing through the sediment ring a few hundred million years after its formation when its radioactive elements have heated all but the crust to the melting point. Although the ring around the Earth and Moon is purely hypothetical, such rings do exist. The asteroid belt, of course, is one. The rings of Saturn are another example. "The surface of the Moon offers a unique opportunity to study the composition of a sediment ring some half billion years after its formation," said Kuiper.

Öpik's Hypothesis

Like Kuiper, Ernst J. Öpik, an astrophysicist of Armagh Observatory, Northern Ireland, held that the density of the large craters on the

Moon was too great to be accounted for entirely by collisions of asteroidal or cometary bodies over time.[16] In 1961 Öpik, then on leave at the University of Maryland, proposed that major crater formation took place early in the Moon's history, when it was hardly more than 30,000 to 50,000 kilometers from the Earth. Öpik theorized that there had been a period of intense bombardment shortly after the dawn of the Solar System. This process represented the closing stage of the building up of the Moon. Öpik suggested that craters and maria must have been formed continuously during the bombardment process, only to be buried under layers of later accretion. It was the last stage of accretion that gave the Moon its present appearance.

The idea that the bombardment occurred when the Earth and Moon were close was supported, Öpik contended, by the observation that the major impact craters were deformed as a result of tidal distortion of a nearby Earth mass.

Öpik considered two sources for the bombarding bodies: Sunbound objects in asteroidal orbits in the vicinity of the Earth's orbit and Earthbound fragments in a cloud or ring orbiting the Earth–Moon system (the sediment ring, again). Like Gilbert and Kuiper, Öpik preferred the sediment-ring explanation.

Öpik placed the Moon at the start of its accretion about 37,000 kilometers from the Earth. If, as he assumed, it was receding from the Earth at the rate of 2,000 kilometers in 80 years (25 kilometers a year), "the whole process of formation of the Moon could have taken place indeed within the bounds of a limited ring of diffuse matter and debris circling the Earth at the aforesaid distance."

Unlike Kuiper or Gilbert, Öpik considered the final bombardment of very large objects the end of a continuous process and not an event that happened only several hundred million years after the Moon formed.

However, if the bombardment was separated in time from the initial accretion process, three successive sources of lunar building blocks would have to be considered, he reasoned: (1) Sunbound material that produced the main body of the Moon; (2) an Earthbound ring that produced the continental craters when the Moon passed through it as it spiralled outward, away from the Earth, after its capture; and (3) asteroidal and cometary bodies from interplanetary space that produced the present craters in the maria.

From Sunbound material (in the Sun's orbit, rather than the Earth's), the Moon would have accreted as an independent planet, as the Earth did, over a period of several hundred million years. Then, 10 to 100 million years after accretion the Moon would have been captured.

For Öpik, the systematic ellipticity of the craters was a powerful clue that the Moon had been formed near the Earth and thus most probably from an Earthbound sedimentary ring. Only when the preferential ellipticities are discounted as due to other causes (such as angle of impact) or to chance, he said, is it possible to accept the alternative of the Moon having accreted from interplanetary material orbiting the Sun.

George Darwin's Moon

The idea that the Moon had fissioned from the Earth was proposed nearly a century ago by Sir George H. Darwin (1845–1912), second son of Charles Darwin. Sir George became Plumian Professor of Astronomy and Experimental Philosophy at Cambridge University in 1883.

His fission hypothesis grew out of his study of tidal interactions between the Earth and the Moon, a subject not only of concern in astronomy but also of enormous practical value to maritime commerce. If a theory making it possible to predict tides anywhere in the world could be developed, its commercial value would be incalculable. Darwin did not succeed in accomplishing this feat, nor has anyone else, but he produced a theory for the origin of the Moon that is still debated.

Darwin first concluded that in the beginning the Earth was molten and rotating rapidly—once every 3 to 5 hours.[17] "The fact," he said, "that the Earth, the Moon, and the planets are all nearly spherical proves that in early times they were molten and plastic and assumed their present round shape under the influence of gravitation."

However, he theorized, this fast rate of rotation was gradually reduced by the friction of tides raised by the Sun on the primitive, molten globe. As the rotation rate slowed slightly, the Sun's tidal effect increased creating oscillations that, combined with the still rapid rotation, "shook the planet to pieces and . . . huge fragments were detached which ultimately became our Moon."

Darwin admitted that "there is nothing to tell us whether the theory affords the true explanation of the birth of the Moon and I say it is only wild speculation, incapable of verification." However, he added, "the truth or falsity of this speculation does not militate against the acceptance of the general theory of tidal friction, which throws much light on the history of the Earth and Moon and correlates the length of our present day and month."

The mechanics of tidal friction are based on mutual gravitational effects of bodies in space. The gravitational effect of the Moon on the Earth raises tides on the Earth. Conversely, the Earth raises tides on the Moon.

The friction effect can be visualized in several ways. One of the least complicated is to think of a tidal bulge raised in the seas. As the Moon passes overhead, it draws the bulge across the world ocean. The moving water mass scrapes the sea bed and pushes up on shore against the direction of the Earth's rotation. The friction thus generated acts as a brake on the turning Earth.

At the same time, the tidal bulge accelerates the Moon's orbital motion around the Earth. How? Perhaps the bulge might be considered a moving concentration of mass that exerts an increased amount of gravitational force on the Moon and ever so slightly speeds up its orbital motion. The acceleration of the Moon's orbital motion causes the Moon to move farther away. The rate of recession is presently estimated at 3 centimeters a year.

The tidal effect has been visualized in other ways. One is that tidal friction distorts the bulge as the Moon drags it around the Earth, moving the bulge slightly off center from the Earth–Moon axis. The distortion results in a torque that slows the Earth's rotation and accelerates the Moon's orbital velocity.

The rotation of the Moon, presumably the same as that of the primitive Earth at the time of supposed fission, was slowed by the tidal friction of Earth's gravity, so that the Moon's periods of axial rotation and of revolution around the Earth are now the same—27.322 days.

It was Darwin's contention that both the day and month are lengthening because of the friction effect. However, the rates of increase in the day and the month are so small that it was not possible to "determine them with any approach to accuracy."

Early records of eclipses observed 3000 years ago show that although there has been no great change in the Earth's rotation or in the Moon's motion since then, according to Darwin "there is . . . a small, outstanding discrepancy which indicates that there has been some change." He added, "In this way it is known that within historical times the retardation of the Earth's rotation and recession of the Moon have been at any rate very slow."*

The ultimate outcome of tidal interaction between Earth and Moon, he said, is a day and a month of the same length—about 55 of our

---

* Estimates of the increase in the length of the day as a result of the slowing of the Earth's rotation range from 1 second in 120,000 years to 1 second in 600,000 years. Laser ranging from the Earth to the Moon, now possible with the emplacement of laser reflectors on the lunar surface, may provide direct evidence of changes in both the Earth's rate of rotation and the Moons' velocity of revolution and recession.

present days. The periods of the Earth's rotation and the Moon's revolution will be identical in the far future, he predicted.

Darwin speculated that in the distant past this situation may also have occurred with the Moon much closer to the Earth, "when the Moon and the Earth went round at the same rate, as though fastened together by a rigid bar." In this early time, he calculated, the periods of the Earth's rotation and the Moon's revolution were 5 hours and 36 minutes. In this state, according to the calculation, the center of the Moon would have been only 6000 miles from the surface of the Earth. The minimum amount of time since the existence of this state, according to Darwin's calculations, was 54 million years.

This led Darwin to "the inevitable conclusion that the Moon and the Earth at one time formed parts of a common mass" and that the Earth–Moon protoplanet broke into two masses, nearly in contact and rotating as parts of a rigid body.[18]

Darwin reasoned that something must have occurred to slow the Moon's period of revolution so that it became longer than the short Earth day. Tidal friction would then operate to retard the Earth's rotation and accelerate the Moon's orbital velocity so that it spiraled outward from the Earth—as it presumably still is doing.

These ideas about the early relationship between the Earth and Moon stretched the state of the art of Earth–Moon dynamics about as far as it would go in the absence of direct measurement and experimentation. Darwin's ideas profoundly influenced thinking in astronomy and astrophysics. The American astronomer, William H. Pickering (1858–1938) suggested that the Pacific Ocean basin was the scar left by the Moon's breakaway—the shock of which, Pickering added, cracked the Earth's crust to form the continents.

Although advances in geophysics dismissed the Pacific as the fission site, the concept of fission itself, especially at a very early stage, has not been disestablished—or established, for that matter.

A persistent advocate of the fission theory has been John O'Keefe, Assistant Chief of the Laboratory for Theoretical Studies, Goddard Space Flight Center of the National Aeronautics and Space Administration. The average density of the Moon, 3.34 grams per cubic centimeter, approximates that of the Earth's mantle and of the silicate part of stony meteorites. The average density of the Earth is 5.5 grams per cubic centimeter, and the difference is attributed to the Earth's iron core. If the Moon spun off the mantle of the Earth, it would not have an iron core—and the density figure indicated the probability that it did not. O'Keefe pointed out that if the fission occurred at all, it did so after the Earth's core was formed.[19]

Early Moon probes indicated that the Moon had no magnetic field comparable to that of Earth. Since a magnetic field is associated with an iron core, lack of such a field around the Moon was further indication that it had no iron core.

If the Moon had fissioned from Earth, it would also be deficient in siderophile (iron-loving) elements, such as nickel, gold, and platinum. They would have followed the iron down to the Earth's core, and they would be as rare on the Moon as they are on the Earth.

On the other hand, if the Moon was a captured planet, it would show a different pattern of iron and siderophile distribution in its outer parts than the Earth does, O'Keefe reasoned. For, lacking an apparent iron core, the Moon's iron and siderophiles would be still distributed in the upper parts.

For this reason, the necessity of getting rock samples from the Moon for analysis seemed critical in determining its origin. The possibility of finding gold and platinum in the lunar hills was fascinating, but peripheral. Lunar scientists were prospecting for knowledge. "Returned lunar samples may give us for the first time a real basis for theories about the origin of the Moon," said O'Keefe.

The Saturnian Ring

This prospect was exciting, especially because the questions of origin and evolution of the Moon were tied together and intertwined with the formation of the Solar System. The Saturnian-ring theory of G. K. Gilbert, asserting that the Moon condensed from a ring of debris around the Earth, had a number of adherents. A. E. Ringwood of the Australian National University proposed that questions about the composition of the lunar crust might be illuminated by supposing that the Moon accumulated from a ring of sediments around the Earth, rather than directly from the solar nebula.[20] Furthermore, Ringwood suggested, accretion may have taken place in a relatively short time—about 10,000 years.

The condensation theory had several difficulties. How did aggregations form in a nebula? In space, radiation pressure could break up bodies less than a few centimeters in size. Urey pointed out that meteorites, asteroids, and comets tend to break up rather than coalesce. Objects acquiring torque or spin from radiation pressure would break up.

If the Moon is at least the product of the fission of a planet, said O'Keefe, "then there will be a great temptation . . . to think of the Solar

System in terms of planetary fission rather than in terms of accretion from a Solar Nebula."[21] Was Pluto, he added, an escaped satellite of Neptune? Was it formed by fission from Neptune? Was Mars the product of fission by a larger body? These were some of the questions raised by O'Keefe, Urey, and others on the eve of the launch of Apollo 11 in 1969.

Return to Genesis

Project Apollo offered the first opportunity of gathering information directly on the Moon. As the project matured in the 1960s, lunar theorists pinned their hopes on the prospect that landings would produce evidence resolving all speculations, arguments, and questions. Manned exploration, they hoped, would uncover the creation. Exploring the Moon was a return to Genesis.

These expectations percolated through a Summer Study of Space Research by the Space Science Board of the National Academy of Sciences–National Research Council in 1965 at Woods Hole, Massachusetts. It was followed by a Conference on Lunar Exploration and Science at the Lawrence High School in Falmouth, Massachusetts.

The Space Science Board Study was chaired by Harry H. Hess, the innovative and influential geophysicist who had played a prominent role in the IGY and who had been one of the promoters of the Mohole Project (an experiment designed to drill a hole through the Earth's crust down to the mantle for samples). The Board Study defined the major questions of exploring the Moon as the structure of its interior and the way it was formed, the composition and structure of its surface, and the history of its evolution.

"A major objective of the geological exploration of the Moon is the development of perspective in viewing our own planet and the solar system," the Board reported. "We are engaged not only in the exploration of space but also in the exploration of time. Key to this perspective in time is the stratigraphic sequence—the order in which deposits of the past were laid down. The difficulty with the terrestrial record . . . is that active mountain building, erosion and sedimentation have destroyed any recognizable remnant of the primordial Earth. At present, we know almost nothing concrete about the first billion years of the Earth's history . . . the surface of the Moon may be one of the few places where a very early stratigraphic record is preserved and decipherable."

The question of the Moon's origin illustrated the relevance of lunar history to the understanding of the Earth, the Board reported. If the

Moon was captured by the Earth, violent disturbances would have occurred on both planets. These would be more recognizable on the lunar
surface than on the much faster changing surface of the Earth. They
might provide clues to identifying such a catastrophic event. There may
have been a rain of fragmental material on the Earth.

If the Moon was formed by fission, its present composition provides
evidence on the degree of the differentiation that has occurred on Earth
since then. If organic material is found on the Moon, it would provide
clues to the early evolutionary stages of life on the Earth. If the Moon
was formed by independent condensation from a nebular mass, its composition should furnish clues to the chemical differentiation mechanism
operating during the formation of the Earth–Moon system.

The Summer Conference on Lunar Exploration and Science then
proceeded to spell out how these objectives could be realized in actual
reconnaissance and then surface exploration.

The Conference envisioned a 10-year program of exploration through
1974, with one or two manned missions to the lunar surface per year.
Since many of the surface expeditions would require two flights each,
three to five Apollo/Saturn 5 vehicles would be needed each year. The
Conference thus contemplated a landing program involving thirty to
fifty vehicle systems.

The landings, it was supposed, would begin in 1969, following the
selection of landing sites by low altitude, orbiting camera spacecraft
during the period 1966–1967. The Conference prospectus emphasized
that "all experiments should be designed to conserve the astronauts'
time, the most valuable scientific commodity on the early mission." The
scientists had been warned by Astronaut R. Walter Cunningham, a test
pilot of great experience, that men on the surface of the Moon would be
working under severe constraints. The supply of oxygen was limited,
both for breathing and for pressurizing the space suit. It would have to
be carried on the astronaut's back in a portable life support system
(PLSS) that fitted on the suit as a backpack. The suit itself, as developed by 1965, was not the best garment to work in, said Cunningham.
It restricted mobility considerably and work had to be done against it.

The initial excursion of the astronauts on the lunar surface would be
limited to 3 hours. Cunningham explained this was based on an estimated oxygen use. The time limit curtailed the radius of exploration,
which was affected by the walking velocity and the amount of leakage
of oxygen from the suit. Cunningham said the radius of a walking man
on the Moon was believed to be about half a mile. Even with a roving
vehicle, these constraints could not be avoided. For the vehicle could
not be driven any farther from the landing spacecraft than the astronaut

could walk in the time remaining of his oxygen supply. Otherwise, the astronaut risked death if the vehicle broke down or bogged down in the lunar soil.

Exploration Priorities

First priority in the surface program was the return of dirt and rock samples to Earth. Second priority was the setting up of scientific instruments that would measure the intensity of moonquakes, meteorite impacts, heat flow, magnetic fields, atmospheric gases, and the flux of solar wind particles.

A lunar exploration tool kit was designed. It consisted of 10-pound capacity sample containers, made of stainless steel or Teflon; a "space-hardened" rock hammer; a rubber mallet; a sun compass; and a stereoscopic camera.

Experimental instruments included a passive seismograph weighing 25 pounds to record the quakes and meteorite hits; a 13-pound magnetometer to measure whatever magnetic field the moon might have and its variation; a 15-pound device to record heat flow; an active seismograph experiment weighing 7 to 20 pounds to measure the effects of man-made seismic waves to be generated by explosives or by the impact of man-made objects on the ground; an atmosphere detector, about 10 pounds; and a micrometeoroid detector, about 15 pounds. Later, a gravimeter would be introduced into the experimental package to record changes in lunar gravity and to pick up, possibly, the mysterious "gravity waves" that may exist in the universe.

The returned lunar samples would be first analyzed at a Lunar Receiving Laboratory, to be built at the Manned Spacecraft Center at Houston before being distributed to university investigators. The Apollo command module would remain in lunar orbit with one man aboard while the other two were on the surface. The command module pilot would carry out orbital experiments, such as high-resolution photo mapping and radiometric studies of the surface.

After several initial landings had established the efficacy of the vehicle system and the modus operandi, a second phase would begin. Phase II was to be carried out by the Apollo extension system (AES).

In AES, the orbital experiments would be increased by the addition of such remote sensing instruments as ultraviolet, infrared, and radar imagers; mapping cameras; ion and neutral mass spectrometers; a solar wind detector; cosmic-ray (solar and galactic) detectors; and an instrument to detect the presence of hydrogen in the surface—a clue to the possible existence of water. The orbital science package would include also a magnetometer, gravity gradiometer, and a probe to measure the electromagnetic properties of the surface and subsurface.

Surface AES experiments included five or six missions with astronauts staying up to 14 days and making traverses of 15 kilometers (9 miles). At least 200 to 250 kilograms (450 to 600 pounds) of dirt and rock would be returned from these missions. Astronauts would make studies from which the construction of a lunar base could be designed.

Travel on the surface would be accomplished by hiking or riding in a vehicle, the LSSM (lunar scientific survey module) capable of carrying two space-suited passengers and 600 pounds of equipment and of covering 8 to 15 kilometers. Also envisioned was a lunar flying vehicle (LFV) capable of lifting 308 pounds of payload 15 kilometers.

The surface vehicle would carry a 1-inch drill that could bore a hole 10 feet deep into rubble or solid rock. The drill could be used to take core samples. These would be extracted by sinking a hollow tube into the ground and pulling it back up. The cylindrical mass of dirt in the tube is the core. The holes could be used also for the insertion of temperature probes to record heat flow from the depths of the Moon.

A third phase of the manned exploration of the surface, called the post-AES program, was sketched. It projected long-distance traverses up to 800 kilometers and a fixed site for investigations lasting 2 months to a year. It envisioned a large, laboratory vehicle, a species of lunar "camper," that would provide transportation and shelter for geologists on long-range field trips. Equatorial traverses up to 800 kilometers or more would be possible in such a vehicle, with a crew of three men.

Finally, a lunar base would be constructed—a complex of surface buildings in which investigators could live for periods up to a year.

The timetable was explicit, if somewhat optimistic. The first landings were to be made in 1969. They would be followed by a 10-year period of geological exploration. The early Apollo and Apollo extension mission would be carried out in the 1969–1974 period. The post-AES mission would begin in 1975 and continue at the rate of at least one a year until 1980.

"An exploratory program of this scope should provide first order answers to most of the major questions that can now be asked about the Moon," the Study concluded.

So the exploration of the Moon was conceived, with great expectations.

Notes

1. *Proceedings of the American Philosophical Society*, Vol. 12, 1892.
2. Whipple, F. L., *Earth, Moon and Planets*, 3rd ed. Harvard University Press, Cambridge, Mass., 1968.
3. Ibid.
4. Baldwin, Ralph B., *Face of the Moon*. University of Chicago Press, Chicago, 1949.
5. Urey, Harold, "Origin and History of the Moon," *Bulletin of the Atomic Scientist*, September 1969.
6. Baldwin, Ralph B., *Measure of the Moon*. University of Chicago Press, Chicago, 1963.
7. Moore, Patrick, *Survey of the Moon*. Norton, New York, 1963.
8. Ibid.
9. Green, Jack, "Lunar Exploration and Survival," a paper by the Advanced Research Laboratory, Douglas Aircraft Co., 1966.
10. *Astronautics and Aeronautics*, NASA, October 11, 1968.
11. Saari, J. M., and Shorthill, R. W., "Physics of the Moon, Review of Lunar Infrared Observations," a paper presented at the American Astronautical Society, American Association for the Advancement of Science, Washington, D.C., December 1966.
12. Urey, Harold, "Origin and History of the Moon," *Bulletin of the Atomic Scientists*, September 1969.
13. Ibid.
14. Ibid.
15. Kuiper, G. P., "Origins of the Moon," *Proceedings of the National Academy of Sciences*, Vol. 40, 1954.
16. Opik, E. J., "Tidal Deformations and Origin of the Moon," *Astronomical Journal*, Vol. 66, No. 2, March 1961.
17. Darwin, G. H., *The Tides and Kindred Phenomena in the Solar System*, 3rd ed. W. H. Freeman, San Francisco.
18. Scientific Papers by Sir George Darwin, Vol. 5, Biographical Memoirs by Sir Francis Darwin and Professor E. W. Brown. Cambridge University Press, Cambridge, England, 1916.
19. O'Keefe, J., "Theory of Lunar Formation," *Bulletin of the Atomic Scientists*, September 1969.
20. Ringwood, A. E., Third Lunar Science Conference, January 10–13, 1972.
21. O'Keefe, J., *Bulletin of the Atomic Scientists*, September 1969.

3

Reconnaissance

With the advent of Project Apollo, human expansion into the Solar System began to move out of the pages of science fiction and into the pages of history. There was a romantic but abiding faith among many people in the national space program that it represented an evolutionary development; the manned exploration of the Moon pointed to a means of human survival on other bodies in space in the event that the terrestrial biosphere was destroyed by cosmic forces or by human perversity. The faith was especially strong among the German-born rocket engineers from Peenemunde, who had been attracted to rocketry in the period between the two World Wars by youthful visions of interplanetary travel. The Saturn 5 Moon rocket essentially was their invention, derived from the V-2. They and many native-born colleagues in the Apollo program conceived of a man-on-the-Moon as the first step in the investigation of the Solar System by human, rather than by robot, explorers. It was the first step in realizing Tsiolkovsky's prophecy, made in the last decade of the nineteenth century, that mankind would settle other planets. How and when this might be done were as vague at the outset of Apollo as in the previous century, but the success of the Saturn–Apollo transportation system lent engineering credibility to the idea that a space-faring species might find another roost in a galaxy filled with stars and, presumably, planets if catastrophe overtook its home world. Apollo greatly enhanced the scenario by moving it out of the realm of fantasy and into the realm of technological possibility. If *Homo sapiens* could colonize regions beyond the Earth, the species might escape the fate that had extinguished its evolutionary precursors and untold numbers of other species. It might acquire relative immortality through technological, rather than biological, adaptation. Perhaps other species in the galaxy had done this long ago.

Such an idea acquired an aura of credibility with the awesome sight of Apollo 11 poised for launch, like a giant finger pointing starward on Cape Kennedy, Florida during July 1969. Only 8 years had passed since President John F. Kennedy had called for the manned lunar landing as a national goal, in a period when aerospace technology could barely boost one man into a low Earth orbit.

In contrast to the engineering challenges of Apollo and the vision behind it were the objectives of the scientists—geologists, geochemists, petrologists, astronomers, and astrophysicists. Another kind of romanticism pervaded their interests. Many of the principal investigators in the lunar science program shared a common faith that the Moon would turn out to be the key to the origin of the Earth and the Solar System. They reasoned: If the Moon were as dead as it appeared to be (in spite of intermittent glows and luminosities people had claimed to have seen on its surface for centuries), its geology and chemistry might reveal the nature of Earth as it was near the time of its formation 4 to 4.5 billion years ago. The first billion years of Earth history had been obliterated, either by meteorites or cooked away by the plutonic fires of the primeval planet. Perhaps they could be found on the less active Moon.

To realize these objectives, the scientists did not require the costly transportation of astronauts to the Moon. Automata that could collect and return samples of lunar rocks and soil and relay data from chemical, seismic, and radiological observations might do just as well. The Russians had been moving in this direction since 1967 and by the time Apollo 11 reached the pad in the spring of 1969, they had created a small, interplanetary shovel that would land on the Moon, scoop up 500 grams of dirt and fly it back to Earth for analysis. Only bad luck prevented this bargain-basement approach to lunar geology from upstaging the triumph of Apollo 11. Luna 15 crashed into the Sea of Crises, but it, too, was a beginning—a first step in another mode of interplanetary investigation that the United States eventually would be required to adopt out of sheer, economic necessity.

Inevitably, there was a conflict between engineering and scientific goals in Apollo. It was conspicuous in the selection of flight crews. Of the twenty-one astronauts manning the seven landing missions to the Moon, only one was a scientist who had been trained by NASA as a pilot. All the rest were test pilots who had received varying degrees of scientific training. At the Manned Spacecraft Center in Houston, Texas, the rationale of the flight directorate was crew performance first, scientific performance second. Throughout the entire Apollo program,

NASA was primarily concerned with the development of a vehicle system that would transport men and some scientific equipment to the Moon and back. From this point of view, the scientific experiments were hitchhikers. So were scientists. The fact that the Saturn–Apollo system, which had been designed primarily to beat the Russians to a manned lunar landing, was too costly to survive a limited series of missions did not influence this policy. NASA went right on testing this system until it reached its inevitable end in 1973. The result was that only one scientist, a 37-year-old geologist named Harrison H. Schmitt, went to the Moon in the Apollo program. Schmitt's assignment to Apollo 17 was made only after considerable pressure had been exerted on NASA and on the space committees of Congress to include a geologist on the last Apollo mission. One should have been included on the first, many scientists believed.

The Cosmogenic Condundrum

Long before Apollo was even thought of, the question of the origin of the Moon was a central one in speculations about the Solar System. In many respects, the Earth–Moon system can be considered a double planet system. How did it come about? Was it unique in galactic terms? Was the Solar System unique? No one had ever detected the presence of a comparable system in the galaxy. Planets orbiting even the nearest stars were indetectable by direct telescopic observation and their existence was merely assumed. Such an assumption rested on a hypothesis of how the Solar System came into being from a gaseous nebula and a critical link in the intellectual process was the Moon.

In theories about the origin of the Moon, the old rivalry between the concepts of catastrophism versus uniformitarianism reappeared. Catastrophism, as I have related, had been laid to rest along with its explanation for a 6000-year-old Earth in the eighteenth century, but it popped up again in the fission theory of lunar origin. As Darwin and Pickering had speculated, the fission theory held that the Moon had spun off a rapidly rotating Earth, leaving the scar of the Pacific Ocean basin and creating the present continents.

Urey, Kuiper, and others suggested that the Moon had been formed elsewhere in the solar nebula and had been seized by the Earth as it swung by. From an exploration viewpoint, this theory was more attractive than the fission hypothesis because it offered scientists another planet, rather than an old piece of the one they lived on, to examine. Also, if the Moon was, indeed, another planet, its evolution, if similar to that of the Earth, might reveal a species of uniformity in planet-forming processes. This could lead quickly to a cosmogenic theory susceptible to confirmation on the Moon.

Another view of the capture theory was proposed in 1966 by S. Fred Singer, a physicist long active in space research. In a paper read at the annual meeting of the American Association for the Advancement of Science that year, he conjectured that the capture of the Moon might have been a catastrophic event that precipitated the outgassing of the atmopshere and the outpouring of the oceans. It might have been the ultimate cause of life.

> . . . The intense stressing and heating which was created through the capture of a massive object, the Earth's Moon, was also responsible for the rapid evolution of an atmosphere and of the oceans through the intensive and rapid defluidization of the surface rocks of the Earth. These events then set the stage for the evolution of life on Earth.

If the origin of the biosphere and the subsequent development of life were unique events, arising out of a chance, catastrophic accident, such as lunar capture, what would the probability be of their occurrence elsewhere in the galaxy? How would it compare with the probability of biologic evolution as a general cosmic process in the evolution of stars, of the elements, of planetary systems? Was there a natural, uniformitarian order in the development of living systems and their environments? Or was there no imaginable order at all? On the cosmogenic scale, was man descended from stars, or was he an accident of negative entrophy in a random universe? Perhaps we would find out on the Moon.

From a uniformitarian point of view, the evolution of the atmosphere and of the oceans could be explained by plutonic processes arising out of the decay of radioactive isotopes. The heat thus emitted boiled vapors and gases out of the rocks. This process did not require the trigger of a wandering Moon.

On the eve of Apollo 11, the fundamental question about the origin and evolution of the Earth and the Moon was whether they represented evolutionary models in a uniformitarian universe or whether they were the unique products of accidental, catastrophic events. Was the Moon the product of plutonic evolution? Had it been cooked, like its primary, the Earth? Or was it simply a clot of primitive solar material or an ancient relict of the Earth's mantle?

Most capture theorists agree that the Moon was seized quite early in the game, perhaps 4 billion or more years ago. One exception has been proposed by the controversial physician turned cosmologist, Dr. Emmanuel Velikovsky. In 1950 he challenged orthodox cosmology with the publication of a book, *Worlds in Collision*, in which he cited evidence from the Old Testament and other ancient texts that turmoil persisted in the inner Solar System until rather recently, in the time of bronze-age man. According to Velikovsky, various catastrophes described in the ancient writings and legends of various peoples were accounts of terrestrial cataclysms, caused by the orbital excursions of Mars and Venus. These planets wandered close enough to Earth to produce massive tidal upheavals in the crust before settling down into their present orbits three to four millennia ago.

What the ancients were trying to tell us, Velikovsky claims, is that the Solar System has not been as stable as orthodox science believes. They were also trying to tell us, he says, that early man remembers a time when the Earth had no Moon in the sky. He cites references to a Moonless Earth by Democritus, Anaxagoras, Aristotle, Plutarch, and Censorinus. These Greek and Roman savants spoke of an ancient people who inhabited Arcadia in Greece, before the Moon arrived, he says. In addition, he cites an interpretation of two Old Testament passages: Job 25:5 "before [there was] a Moon and it did not shine" and Psalms 72:5 "thou wast feared since [the time of] the Sun and before [the time of] the Moon."*

If the Moon appeared in the time of man, Velikovsky reasoned, it must have been captured by the Earth.[1] But most capture theorists reject the notion that capture was so recent. They are certain that the Moon has been around a very long time. How long? Perhaps the Moon, itself, would tell us.

A Close Look

Before a landing on the Moon could be attempted, it was necessary to take a close look at the surface and to select landing sites. Early in the 1960s, as space-flight navigational and maneuvering techniques were evolving in the Mercury and Gemini programs, three lunar reconnaissance projects were being developed. They were the Ranger project of crash landing a camera that would take pictures and radio them back to Earth as it approached the Moon until it crashed; the Surveyor project of soft landing more sophisticated spacecraft that would take photographs after touchdown and radio them back, along with other data on the nature of the lunar soil; and, finally, the Orbiter project, consisting of a series of camera spacecraft in a low orbit around the

---

* This translation from the Hebrew does not agree with that in the standard, King James version of the Old Testament.

Moon photographing large areas for landing sites and geologic information. At the time these projects were in development, no one knew
whether a vehicle could land on the Moon without sinking out of sight
in deep dust or plunging through a weak, brittle, piecrust surface into a
cavern or crevasse. No one had seen lunar surface detail any smaller
than a kilometer in diameter, the maximum telescopic resolution permitted by Earth's atmosphere.

In Project Ranger, two technologies had to be perfected: A radio
camera system that would photograph the Moon as the vehicle approached it very rapidly and transmit the photos back to Earth receivers
and a rocket capable of launching the reconnaissance package all the
way to the Moon. Phototelemetry systems had been developed in military reconnaissance, but rocketry was another problem. Starting August
17, 1958, with Able 1, the United States fired nine lunar probes, with
zero accuracy. Every one missed the Moon. The tenth shot, Ranger 4,
launched April 23, 1962, crashed on the far side of the Moon. The
Russians had better luck. They hit the Moon with Luna 2, on their
second try September 13, 1959, and succeeded in photographing the far
side, a historic event, with Luna 3 the following October.

When NASA's Ranger finally did work, however, it worked beautifully. Ranger 7 took and radioed back the first close-up photos of lunar
surface as it came in for a crash landing on the fan-shaped rays of the
Crater Tycho on August 31, 1964. In all, the Jet Propulsion Laboratory
at Pasadena, California, the NASA facility operated by the California
Institute of Technology, received 4,308 pictures. Ranger 7 was one of
the great triumphs of the space age.

Now, an amazing lunar landscape stood revealed, a scape no one had
been able fully to imagine. The final Ranger 7 photo taken from an
altitude of 1600 feet-seconds before the vehicle struck the ground
showed craters as small as 10 inches in diameter. Lunar seeing had
been enhanced a thousand times. The Ranger photos were as much a
revelation as Galileo's first telescopic glimpse of the Moon had been in
1609, disclosing its mountains and valleys.

The view from Ranger revealed a heavily pitted surface, craters
nested within craters, like a bombarded World War I battlefield in
France. There were small craters on the rims and in the basins of big
ones; old craters virtually erased by impacts that had made new ones.
Curiously, the edges of the craters and of the rocks and the sides of hills
were gently rounded—not sharp, as commonly thought.

This observation upset the deeply embedded notion that the lunar
surface held knife-edge scarps and corners. Without the erosion of wind
and water, which were absent on the Moon, how could it be otherwise?

But it was otherwise. Erosion had been at work on the Moon. The rounding of edges and the sandpapering of sharp corners were the work of some erosive agency, possibly the solar wind or the infall of space dust. The "extraterrestrial look" that artists had imparted to the Moon by means of a jagged rockscape was nowhere to be seen. Instead, lunar investigators were looking at a rolling desert studded with potholes of all sizes.

Ranger 8, which returned 7,137 photos before it crashed in the western side of Mare Tranquillitatis on February 20, 1965, and Ranger 9, which returned 5,814 photos before plunging into the floor of the Crater Alphonsus in the highlands east of Mare Nubium on March 24, 1965, disclosed other surprises. Dimple craters, rilles, and long, sometimes sinuous valleys appeared in these pictures. They suggested isostatic adjustment of a surface that had experienced a transfer of mass, such as a lava flow. Isostasy works like this: If you push in one part of a balloon, another part bulges out. If the skin is fairly rigid, like the rind of a melon, it may crack. Rilles are thought of as cracks. The dimple craters betrayed the existence of voids, or caverns or crevasses, overlain, perhaps, by a thin, brittle crust. They suggested past volcanic activity. In the lunar highlands, Ranger 9 photos picked out "halo" craters, surrounded by material ejected from them. They showed chains of craters that appeared to be volcanic, rather than punched by meteorites. The evidence for volcanism was growing, but Ranger did not reveal enough evidence to indicate that volcanism was a major force in shaping the surface. None of the 17,259 photos radioed back to Earth by the three successful Rangers provided any definite information about the bearing strength of the soil. During this period of lunar reconnaissance, the Grumman Aircraft Engineering Company was building the lunar module on Long Island, New York. The lunar module or LM, was the vehicle in which two Apollo astronauts would descend from the command ship in lunar orbit to the Moon's surface and later return to the command ship by means of rendezvous and docking learned in the Gemini Program. One of the main problems in the LM was the landing gear. Grumman engineers had devised one without knowing for sure what kind of material the vehicle would land on. Sensibly, the company anticipated a touchdown surface with a bearing strength somewhere between that of swamp mud and beach sand.

Surveyor

Ranger was followed by a series of soft-landing automated probes called Surveyors. The technology required for building a vehicle that could fly to the Moon and execute a soft landing at a predetermined site stretched the state of the art in radar altimetry, computers, and rocket engines. The Soviet Union broke through first, landing its 225-pound Luna 9 in Oceanus Procellarum west of the Crater Reiner on February 3, 1966, after four earlier attempts failed. The little vehicle that stood only 2 feet above the ground came down on a hard surface. The ground appeared from photos radioed back to be a crumbled lava bed. This was the first indication that it was possible for a vehicle to land in the maria regions of the Moon—the dark areas, which look like lava flows.

Five months later, after 28 months of delay, the United States Surveyor 1 landed in the southwest part of Oceanus Procellarum on June 2, 1966. After bouncing lightly on the level mare floor, the 5-foot-tall vehicle came to rest and its camera radioed 11,240 pictures to Earth during 2 lunar days (of 14 Earth days each). The vehicle was finally shut down by radio command on January 7, 1967.

The view from Surveyor 1 showed a dark, relatively smooth surface, encircled by rolling hills and low mountains, that formed the rim of an unnamed crater not far from a well-known landmark, the Flamsteed Crater. The landscape appeared to be studded with craters of all sizes, many containing angular rocks ranging from a millimeter to a meter in diameter. In size and distribution the number of smaller craters was similar to that shown by the Ranger photos. However, just as Ranger had increased the view of the lunar surface by three orders of magnitude, compared with telescopic viewing from Earth, so did Surveyor's television camera increase resolution by three orders of magnitude over Ranger.

In Surveyor photos, the surface looked like a freshly ploughed field. It was composed of granular material in which large, coarse blocks were resting. On landing the Surveyor footpads had pushed aside and compressed some of the fragments. The pads of the 640-pound (Earth weight) vehicle had sunk 3 to 8 centimeters into the surface. Their impact had produced clods of soil. The mechanical response of the lunar soil appeared to be similar to that of a damp, fine-grained terrestrial soil, although there was no trace of water on the Moon. Was this soil, analogous to that of Indiana or Illinois cropland, capable of growing anything? Surveyor investigators wondered. The surprising answer was to come later.

The following September 20, Surveyor 2 was launched, but it crashed in the southeast section of the Crater Copernicus when a vernier engine failed. The next year, the space agency succeeded in landing Surveyor 3 in the eastern part of Oceanus Procellarum on April 20, 1967. During the first lunar day, its television camera sent back 6,327 pictures, again depicting the soft, barren landscape, pocked with craters and crowned by rounded hills.

Surveyor 4 failed. Then, on September 10, 1967, Surveyor 5 landed in the southwest part of Mare Tranquillitatis near the top of a crater about 12 meters wide. In addition to its television camera, which returned 19,119 pictures in four lunar days, Surveyor 5 carried a chemical analyzer that was designed to report the relative abundances in the soil of chemical elements between hydrogen and silicon in the periodic table.

The analyzer worked by bombarding the soil with alpha particles (helium nuclei) that were emitted by radioactive curium-242. Some of the particles bounced back when they hit the nuclei of chemical elements in the range of atomic weights between hydrogen and silicon. The energy of the bounce varied with the elements struck, and this was measured by sensitive detectors. Alpha particles that hit an atom of oxygen, for example, bounced back with a different energy than particles striking an atom of aluminum or sodium or magnesium. Each element reflected or scattered the alpha particle in a characteristic way. In addition, some of the bombarded elements emitted protons. These, too, characteristically identified the elements emitting them. The alpha-particle scattering instrument was housed in a gold-plated box 6 inches on a side. It was lowered to the ground on a nylon string automatically after Surveyor 5 had landed. Data recorded by the instrument's sensors were radioed to Earth.

The alpha-particle scattering device was designed at the University of Chicago by a team under the direction of Anthony Turkevich, professor of chemistry in the Enrico Fermi Institute of Nuclear Studies, and James H. Patterson, a research chemist at the Argonne National Laboratory, an Atomic Energy Commission facility southwest of Chicago. Ernest Franzgrote of the Jet Propulsion Laboratory, which had masterminded the Ranger and Surveyor programs, was a co-experimenter on the analyzer.

After 900 minutes of operation on the Moon, the little gold box returned the data that spelled out in the main the chemical composition of the soil on which Surveyor 5 rested. It was a type of basalt, like the most common igneous rocks that form the ocean floors and ridges on Earth.

The discovery that basalt was the dark material of the maria at this one location, and, probably, at all the maria basins, is a landmark in science. And yet it was not unexpected, for the dark floors of the lunar basins looked like basaltic lava flows, even through a pair of field glasses. Now the Moon was beginning to show a terrestrial-style evolution, as many investigators also had expected. The little gold box reported that the chemical composition of the lunar maria surface was 58 percent oxygen, plus or minus 5 percent; 18.5 percent silicon, plus or minus 3 percent; 6.5 percent aluminum, plus or minus 2 percent; and 3 percent magnesium. Surprisingly, 13 percent (plus or minus 3 percent) of the elements heavier than silicon appeared in the data, but the alpha-particle scattering method was not capable of identifying them. However, iron, cobalt, and nickel were estimated at 3 percent of this residue and the heavier elements at not more than 0.5 percent.

The investigators reported:

> The overall analysis indicates that the lunar surface at the Surveyor 5 landing site is a silicate rock similar in composition to materials available on Earth. The results are more comparable to the chemical composition of the continental crust of the Earth than to that in the outer region of the Sun. Both the Moon and the Earth have much less magnesium and aluminum relative to silicon than does the Sun. However, there is less sodium and more atoms of elements heavier than silicon on the Moon than in the Earth's crust. There is also an indication that the silicate content is lower on the Moon than on the continents of Earth. The conclusion indicated by this comparison is that if the Earth and the Moon were originally formed from solar type material, the major geochemical changes to the material at the Surveyor 5 landing site must have been similar to those that occurred to the materials comprising the terrestrial continents.[2]

This result, the investigators added, was obtained on less than 100 square centimeters of material by analysis only of the top few microns in one place on the Moon. Later, they added that the Surveyor 5 sample resembled "most closely a rock of basaltic type, a terrestrial basalt or a meteoritic basaltic achondrite."[3] There was no doubt that the material was the product of chemical processing, like the rocks of Earth.

Confirmation of the alpha-particle scattering experiment findings was provided by a small bar magnet carried on footpad No. 2 of the Surveyor. The Alnico V magnet, 5 centimeters long and 1.2 centimeters wide, was dusted off by a mild blast from one of the spacecraft's vernier engines after landing and then studied closely via the television camera. Material that remained stuck to the magnet was interpreted as having high magnetic permeability, such as iron, magnetite (ferric oxide), or nickel-iron fragments of meteorites. The amount of magnetic material that remained on the magnet after it was dusted by the vernier engine was about as much as would be expected in a powdered basalt. The lunar magnet results, reported experimenters, "are not in disagreement with the laboratory results of impact in a 37 to 50 micron powdered basalt with no addition of iron."[4]

The amount of magnetic material was considerably less than that expected if the surface had been meteoritic.

Implications of Basalt

The discovery of basalt on the Moon was of the order of importance of the discovery of the radiation belts around the Earth. It appeared to be evident that chemical evolution on the Moon had followed a similar uniformitarian pathway to that on the Earth. This meant that all or part of the Moon had been hot enough to have melted the original "ultra-mafic" rock, so that the volatile or lighter elements boiled away, leaving the heavier or refractory elements more concentrated than when the Moon was first formed. The separation of the volatile from the refractory elements, a process called fractionation, has occurred extensively on Earth as a result of internal heating and melting from radioactive sources. It is the process that has produced terrestrial volcanic or plutonic rocks.

With the finding of basalt on the Moon, scientists could infer extensive volcanism and from that they could apply the terrestrial process of fractionation and differentiation to another planet. On Earth, basaltic magma, vomited out of the planet by the rising heat of radioactive decay within, had filled the great basins that became the ocean beds. Since some of the terrestrial basalts appeared only 300 million years old, great outpourings of lavas on Earth had been relatively recent. At least the layers accessible to man appeared to be recent in terms of geologic time. On the Moon, lavas had also flowed into basins that Galileo and his contemporaries interpreted as seas. The plutonic processes were similar. Surveyor investigators exulted:

> It is significant and gratifying that the chemical composition of the lunar material appears to be most like that of a common, terrestrial rock. . . . Apparently the geochemical processes on the Earth do not differ greatly from their lunar counterparts, despite environmental differences in the two bodies.[5]

For the first time there was scientific validity in extrapolating terrestrial geochemical and geologic experience to the analysis of the Moon and the processes that formed it. This in itself was a remarkable advance, suggesting a consistent evolutionary pattern for all terrestrial-type bodies in the Solar System. After accretion from solar nebula material, bodies at least as large as the Moon heated up as a result of the energy emitted by the decay of radioactive isotopes, principally uranium-235, uranium-238, thorium-232, and potassium-40. Rocks melted, fractionated, flowed, cooled, and recrystallized. On Earth, additional processes of erosion, deposition, and consolidation in the oceans had produced sedimentary rocks, which had not appeared on the windless and waterless Moon. Nevertheless, a grand design was beginning to emerge. Confidence grew that further investigation would reveal its true dimensions and that mankind would learn a great truth about the Creation, and learn it soon.

The Regolith

On November 10, 1967, Surveyor 6 landed in the Sinus Medii near the center of the Moon, a short distance from a ridge in a level mare region. It carried the alpha-particle scattering analyzer and magnets in addition to the television camera. During the first lunar day, the camera took 29,952 pictures and radioed them back to Earth through the Deep Space Network receiving stations, which were to play an important communications role in Apollo. The stations were at Goldstone, California, in the desert; Tidbinilla, Australia; Robledo, Spain; and Johannesburg, South Africa. Surveyor 6 returned data for a little more than a month. It was shut down on December 14, 1967.

Again, the alpha-particle scattering experiment reported a predominantly basaltic composition of mare surface material.[6] Commented the experimenters:

> The similarity of the results at the two Surveyor landing sites makes it improbable that the chemical composition found is applicable only to unique places on the Moon. It is much more probable that they are representative of the large portions of the lunar maria that are similar in appearance and in optical and thermal properties.

Then, on January 10, 1968, Surveyor 7 settled down in the bright lunar highlands on the ejecta blanket north of the rim of the huge Crater Tycho. Surveyor 7 landed in a hilly highland region in the Moon's southern hemisphere and returned 21,038 pictures and other data until it was shut down on February 21, 1968. This was the first highland site sampled by a Surveyor and the alpha-particle scattering experiment was expected to report a different chemical composition in the white soil than in the dark maria. It did not disappoint anyone. It reported that the highland sample had about half as much of the iron group elements (titanium through copper) as samples from the lowland maria. The lower iron content of the highland soil explained its higher albedo or brightness. Highland material also appeared to be less dense than mare basalt, as it was weighed and moved about by a soil sampling device, an electrically operated scoop, carried by Surveyor 7 (and also by Surveyor 3).

Surveyor 7 was the finale of that program, the most successful and complex of any lunar investigation attempted up to that time. Scientific teams reported that the lunar soil analyses at both mare and highland sites indicated that the most abundant chemical element on the Moon is oxygen, 57 percent plus or minus 5 percent; the next most abundant is silicon, 20 percent plus or minus 7 percent. These are also the most common elements in the Earth's crust in the same order of abundance. Samples analyzed at the two maria sites by Surveyors 5 and 6 were chemically "almost identical," the teams reported.

Surveyor demonstrated that the Moon was definitely not meteoritic. It was an evolved, or partly evolved, planet.

"The chemical analyses are in strong disagreement with that expected for primordial solar system material," the investigators reported, "whether this be considered condensed solar atmosphere, terrestrial ultrabasic rocks or chondritic meteorites."[7]

The analysis dimmed a long-held theory that the Moon was a source of meteorites. And it dismissed the notion of a lunar origin for tektites, a class of small, glassy objects found in clusters on Earth and once thought to be the splash-out of great meteorite impacts on the lunar surface.

The idea that tektites, which appear to have been fused by the heat of entering or reentering the atmosphere into a rough teardrop shape, came from the Moon was fairly widely believed. The shape of the glassy spherules suggested aerodynamic ablation after they were splashed out of the Moon's gravitational field and then captured by the Earth. Tektites found in Australasia appeared to be related chemically to the surface material at the Surveyor 7 landing site on the rim of Tycho, according to a report from NASA's Ames Research Center at Mountain View, California. A computerized trajectory analysis indicated that the objects could have originated from the meteorite impact that created Tycho. But the alpha-particle scattering analysis that revealed that lunar and terrestrial basalts were chemically similar showed it was just as plausible that the tektites represented splash from terrestrial volcanoes or meteorite impacts.

Beyond the chemical composition of the surface dirt, the Surveyors provided additional evidence of a history of lunar volcanism. Fragments near Surveyor 7 in the highlands and near Surveyors 1 and 3 in the maria were seen to have deep surface pits, ranging in size from a fraction of a millimeter to a centimeter. The pits seemed to be vesicles, or cavities, made by gas rising out of a molten magma as the rocks crystallized.

Although the identification of basalt in the mare basins gave circumstantial evidence of volcanism of considerable extent, the theorists who relied on impact to explain the large lunar structures did not believe that they had been unhorsed. Two investigators of the U.S. Geological Survey (USGS), Eugene Shoemaker and E. C. Morris, asserted that "the large majority of craters observed on the lunar surface in all size classes is here interpreted to be of impact origin."[8] That was the predominant process of crater formation; it was the ending, the last inning in the cosmic ball game of planetary accretion. Planets grew this way in their final stages, sweeping up the debris left over in their neighborhood, scooping up an occasional passing clump of rock, perhaps from the so-called asteroid belt, and the force of impact left depressions in the surface—depressions that on an airless and waterless body were slow to disappear. Most of the craters seen by the Surveyor television cameras were secondary ones—craters formed by the impact of rocks splashed out of the big impact craters, according to USGS scientists.

Nevertheless, the volcanic, or apparently volcanic, basalt filling the mare basins had to be accounted for. At some time, the Moon was hot—hot enough to bring rocks to a boil and produce the lithic soup of lava. Somehow the lava had gushed out of the depths of the Moon and flooded the basins, as had the lavas on Earth. For there it was, identified under the alpha-particle scattering instrument, clinging in tiny globs to the Surveyor magnets, a basalt rich in iron.

J. Negus DeWys of the Jet Propulsion Laboratory, who had made the magnet studies, cited pre-Surveyor telescopic scans of the Moon that had shown "hot spots" in and around the edges of craters. He recalled the long history of "transient events" or mysterious glows and fogs on the Moon that terrestrial Moongazers had been reporting for more than three centuries. At least 600 of these "events" had been recorded. They ranged from obscure hazes to lightninglike streaks. About a third emanated from a single area—the Herodotus-Schröter Valley region in the Aristarchus Hills on the western side of the Moon north of Oceanus Procellarum. DeWys was certain that such repetitive activity suggested volcanic outgassing. He cited also the observations of N. Kozyrev, the Russian astronomer, who reported gas emissions from the central peak of the Crater Alphonsus. He reasoned that basalt would be expected over most of the lunar surface if widespread volcanism had occurred and if the craters were the product of internal, rather than external, processes. If the surface material and the craters were the results of impacts or of accretion, however, the basaltic floor of the maria would show evidence of meteoritic composition, that is, some addition of meteoritic iron would show up. But it did not. The abundances of iron found in the alpha-particle scattering experiment were those of basalt. The magnet test seemed to confirm this.

The volcanic versus impact controversy was no nearer resolution at the end of the Surveyor program than at the beginning. But a great deal had been learned about the Moon, including the discovery that it was chemically more like the Earth than many investigators had expected and (also) somewhat different. The basalt had been cooked with slightly different ingredients, but the similarities were greater than the differences.

From a practical point of view the bearing strength of the surface material, a conglomeration of fine particles and large fragments the lunar investigarors called "regolith," was of first importance. Could it bear the weight of the lunar module, which was considerably heavier than the Surveyors? The Surveyor surface samplers and footpad penetration observations by television cameras showed the soil was amply firm enough to bear the weight of the LM, about 2,900 pounds on the Moon with the ascent stage fully fueled and the descent stage fuel tank empty. There were no indications of quicksand or of fairy-castle structures into which a vehicle might sink. Such fears, engendered by the Ranger photos, were laid to rest.

The regolith, or topmost soil of the Moon, appeared to be a layer of fragmental debris with low cohesion lying over a more coherent substratum. It was found everywhere and it varied in thickness from a few centimeters to an estimated 10 meters ($32\frac{1}{2}$ feet).

It seemed likely that the regolith had been formed by repetitive bombardment, which had also produced the craters. About 90 percent of it at the Surveyor sites consisted of fragments smaller than a millimeter. At the Surveyor 6 site, on the plains of Sinus Medii (Central Bay), this fine material appeared to be 10 meters thick. As the Cornell University astronomer Thomas Gold had forecast, there was indeed dust on the Moon; some of it was deep. But the dust was cohesive enough to exhibit the mechanical strength of finely textured soil.

The final Surveyor Report expressed doubt that any primitive lunar material had been preserved in view of the generally pulverized condition of the soil. The rocks that could be seen appeared to have been thrown out of craters.

If the basaltic layer detected by the Surveyors was the result of fractionation by heating, the Moon probably had been modified through cooking since it condensed from the solar nebula. That seemed to make sense to a majority of Surveyor scientists. It seemed to rule out any hope that the primeval material of which the planets first were made would be found on the Moon. "A differentiated Moon would imply that the (larger) terrestrial planets also are likely to have differentiated," they said.[9] That might be a safe bet, although Mars might tell another story. We shall see later—for Mars is involved in the lunar deduction, as a witness, one might say.

Gold's Dissent

Gold dissented from the conclusion that volcanism and fractionation had played important roles in the evolution of the lunar surface. Not only did the Moon's apparent high structural strength deny the likelihood of a molten interior, but it had sustained much less horizontal deformation than any widespread volcanism would imply. Where were the folded mountain chains and distortions of high ground that would be expected if large volumes of mass had been displaced by mighty outpourings of lava?, Gold asked.[10] These did not appear nor were there widespread examples of stratification, even on the slopes of large, fresh craters.* If the Surveyor program had demonstrated anything, said Gold, it was that the lunar surface is composed in general "of very fine, slightly cohesive rock powder."

To which white-haired Harold Urey shook his head and responded by repeating his dictum: The Moon is peculiar in that new discoveries about it can be used by each investigator to confirm his own ideas.

Lunar Orbiter

While Surveyor was in progress, NASA sent a series of camera spacecraft into orbit around to the Moon to take panoramic and close-up photos of the surface. This program, called Lunar Orbiter, produced arrays of beautiful and fantastic photos of the lunar world at resolutions 10 times better than those possible with telescopes on Earth. Starting with the launch of Lunar Orbiter 1 from Cape Kennedy on August 10, 1966 and continuing until Lunar Orbiter 5 was sent crashing into the Moon on January 31, 1968, the five orbiting camera craft photographed 99 percent of the Moon, so that virtually the entire surface could be studied. Eight possible landing sites for Apollo were selected from the Orbiter photos. The program was skillfully handled by NASA's Office of Space and Science and Applications. It was managed by the agency's

---

* But these were to appear later.

Langley Research Center at Hampton, Virginia, with the Boeing Company, Seattle, as prime spacecraft contractor. Like Surveyor, Orbiter was a resounding success, technically and scientifically.

Interpreting planetary photography is a specialty of astronomers and Gold was no amateur at it. He saw in the Orbiter pictures little new information substantiating volcanism as a major effect in the molding of the Moon. But geologists disagreed. They saw volcanic features everywhere. The dispute was reminiscent of the nineteenth-century feud between the astronomers and physicists on the one side and the geologists on the other over the means of interpreting the age of the Earth. The discovery of radioactivity and the measurement of rate of decay had confirmed the geologists in that battle, but the melee surrounding the Moon was just starting in the Surveyor–Orbiter period.

The Orbiter program had the primary mission of locating suitable landing sites for Apollo; however, it accomplished much more than that by providing valuable photographic and orbital data. Site selection became a secondary result of the program. Orbiter served as a prototype of planetary photo mapping, later to be carried out brilliantly by Mariner 9 orbiting Mars.

Each of the five Orbiter spacecraft weighed 850 pounds and was 5 feet in diameter and 5½ feet long. Each carried a 150-pound photographic system equipped with a wide-angle and a telephoto camera that viewed the lunar landscape through a protective quartz window. The cameras were loaded with a 260-foot roll of 70-millimeter Kodak special high definition aerial film (Type SO-243) for 212 dual exposure frames. It was necessary to shield the spool from ionizing radiation in space, especially from solar flares. The camera system was equipped with an image compensation device to prevent blurring caused by the rapid passage of the spacecraft across the lunar surface. At perilune, the lowest point in its orbit, the camera was traveling 4500 miles an hour relative to the ground it was photographing.

Following exposure, the film automatically moved into a storage area and then onto a processing drum. It was mechanically pressed against a Bimat processing web and developed. The film negative, fully processed, was dried and the image then was radioed to the Deep Space Network receiving stations by a transmitter that scanned the film with a beam of light and converted its modulations, which corresponded to light and dark areas on the negative, into electrical currents. These, in turn, modulated a radio signal beam to the Earth receiver where the signal was converted back into a photographic image.

During August 1966, Orbiter 1 passed over and photographed the area of the Flamsteed Ring in southwestern Oceanus Procellarum where

Surveyor 1 had landed on June 2 of that year. Interpreting the Orbiter 1 photos, L. C. Rowan of the U.S. Geological Survey reported that they demonstrated a complex relationship of impact, volcanic, tectonic, and mass-wasting processes in the history of the Moon. Although most of the craters appeared to be of impact origin, Rowan reported, there was evidence of volcanism. "Volcanic activity is widely manifested in both individual structures and large provinces," he said. "Deposition of volcanic material subdues subjacent topography. . . . Cones, craters, rilles and ridges are frequently of volcanic origin. . . ."[11]

One of the most important discoveries by the Orbiter program had nothing to do with photography. Observers at the Jet Propulsion Laboratory noted changes in the flight path of Orbiter 5 that could be explained only by changes in the force of gravity as the vehicle passed over some sections of the Moon. Using radio-tracking data the investigators, William L. Sjogren, Paul M. Muller, and Peter Gottlieb, found that every time Orbiter 5 passed over six maria—Imbrium, Serenitatis, Crisium, Nectaris, Humorum, and Aestatum—it accelerated. The acceleration was caused by an increase in gravitational attraction, reflecting an increase in mass below the flight path. Since the increase was in the low-lying maria, not in mountains, it was evidence there was a concentration of mass buried under the circular basins. The investigators referred to these anomalous masses as mascons—short for mass concentrations.

The initial announcement of this discovery caused a furor among lunar scientists. Then six more mascons were found distorting the flight path of Orbiter 5. They were in Mare Orientale, Mare Smythii, Mare Humboldtianum, Crater Grimaldi, and two unnamed mare areas.

One of the early explanations preferred was that the mascons were the buried meteorites or planetoids that had created the basins in the first place. The idea appealed to impact theorists. The volcanic school had another answer, however. The mascons were not extralunar at all, but simply puddles of congealed lava that had settled into the basins from volcanic outpourings and had remained near the surface.

Hot Moon, Cold Moon

The discovery of a dozen mascons on the Moon provided new ammunition for the defense of the theory (held by Urey and others) that the Moon had accreted cold in the solar nebula. Urey had cited as evidence for this the Moon's irregular shape and its offset center of mass. These distortions could not have existed if the Moon had been hot and plastic in the beginning, for it then would have assumed a more hydrodynamical (spherical) shape. If the Moon had ever been molten, its mass would have reached hydrostatic equilibrium—a uniform distribution of mass, as Jell-O does when it cools in a bowl. The mascons lent support to the cold Moon theory because they could not have been held up so long near the surface had the moon been hot. To support the mascons, the Moon had to be cold. It had to be rigid—as rigid as steel.

On Earth, masses were more evenly distributed throughout the body of the planet by the settling out process called isostatic equilibrium, or isostasy. Mascons would not persist on Earth for the excess mass would soon be distributed through a plastic crust and mantle. Mascon persistence on the Moon indicated it was cold and had been cold since the mascons formed.

On Earth gravitational attraction is greater in mountains because more mass is piled up there and is less in ocean basins where there is less mass, compared with the planetary average. On the Moon, Orbiter 5 demonstrated the opposite. Gravity was higher in the basins. Urey calculated that the positive gravitational anomaly of the mascons amounted to 750 milligals, compared with maximum positive anomalies of 200 milligals on Earth. At three-fourths of a gal (a unit of acceleration named for Galileo),* the lunar mascon anomalies are more than three times the highest on Earth and show a great unevenness in mass distribution, which Urey said could not persist in a body softened or made plastic by volcanic heat.[12]

Urey further calculated that the positive anomaly in Mare Imbrium could be produced by a flat, circular slab of chondritic material four kilometers thick and 670 kilometers (402 miles) in diameter, lying near the surface. Was this a chondritic planetoid that had crashed into the Moon to create the Imbrium Basin? In area, it was enormous—nearly 127,000 square miles. If such a mass 402 miles in diameter existed, it was larger than the state of Arizona. In fact, it was only slightly smaller than the inner mare itself, which is 680 kilometers in diameter.

"Now the existence of these mascons shows that the Moon is rigid," Urey said in 1969. "The outer parts of the Moon for surely several hundred kilometers deep cannot have melted. This does not exclude the possibility that the center or the deep interior is partly melted. But . . . one should wonder if lava flows are possible on an object with such a gross, non-equilibrium situation. How do we get basalt on a Moon that is cold and rigid?"[13]

The obvious answer to Urey's question was the suggestion of Gold—

---

* 1 gal of acceleration is 1 centimeter per second per second. The acceleration of gravity at the surface of the Earth is 10 meters per second per second, or 1000 gals.

that the basaltic material was fractionated (cooked) in the solar nebula before the Moon formed and became part of the Moon through accretion. Urey and others, including Gordon J. F. MacDonald, an astrophysicist then at the University of Arizona (1969), considered that the mascons were indeed the residues of planetesimals that had fallen into the Moon and had created the great ringed maria. The impacting bodies were assumed to have a higher density than the average for the Moon and to have hit the Moon at a rather low velocity.

A rival theory of mascons, advanced by the volcanists, held that these concentrations represented a lateral transfer of mass in or on the Moon. "A necessary prelude to this process is that the Moon differentiated (or acquired) a crust of lower density than the interior," observed William M. Kaula, professor of geophysics at the University of California, Los Angeles.[14] "So that—when an infalling body created the preringed mare crater and threw out a lot of material, the mare remained a topographic depression after the crater had been isostatically compensated by a 'root' of denser, subcrustal material. The mass excess was then constituted by the sediments which filled the basins."

It was also proposed that the excess mass was transferred inside the Moon by convection, the process by which mass is transferred laterally on Earth. Source of the convection pressure, Kaula suggested, was thermal contraction, "as would have occurred if the outermost layer of the Moon was hot enough to differentiate a crust while the interior was cooler." The greatest outpourings of lava would have occurred in the ringed maria, it was reasoned, because they were the weakest parts of the lunar crust.

The mascons were only one of the discoveries on the Moon from which advocates of both the hot and cold Moon theories could draw aid and comfort. Before the mascons were discovered, the hot Moon theory had been gaining support, especially from Ranger and Surveyor indications of volcanism. The hot Moon had been proposed in 1957 by Donald H. Menzel of the Smithsonian Astrophysical Observatory in a theory of the origin of the Solar System. Recalling it in 1966, at an American Astronautical Society Symposium, Menzel said he thought of a primordial nebula with about 10 times the mass of the Sun. Contracting because of gravitational forces, the nebula evolved into a huge star, surrounded by gas. Under the influence of magnetic forces developing within it, the star was squeezed into a flattened disk, with a radius equalling the distance between the Sun and Jupiter. As the giant disk rotated, instabilities developed. Successive rings of material became detached from the disk and condensed into planets.

Menzel then visualized a stage of Solar System development in which the Moon was "actually a boiling, bubbling, liquid mass." Although he preferred the idea that the heat was primordial in origin, "we cannot dismiss the idea that some form of radioactivity may have liquefied the Moon."

The Smithsonian astronomer proposed that the cooling lunar mass would form into an ever thickening crust, afloat on the heavier liquid. Tides caused by the Earth would cause this crust to crack. Internal lava would flow out onto the surface, filling the basins. Photographs of the Crater Wargentin, which showed it filled nearly to the top, illustrated how this might have happened.[15]

At the same AAS Symposium in 1966, cited in several instances here because it so well represented the views of the pre-Apollo period of lunar investigation, another scientist, S. Miyamoto, expressed another view. He suggested that a "simple and natural interpretation" of the features shown by the Ranger and Surveyor photos was a quiet solidification of fluid lava. Miyamoto also suggested that the flashes and streaks of brightness seen on the Moon by hundreds of people and reported persistently since the year 1587* indicated boundary phenomena betweend maria and highland zones. The terrae or highland regions and the maria lowlands may represent different crustal materials, analogous to the continental blocks and ocean basins of Earth, he said. This idea was to flower dramatically after men landed on the Moon.

Urey could not accept the idea of a hot Moon. In his view, the mascons ruled it out. ". . . I find it difficult to understand how these objects could have been supported by the Moon if it were at the melting point throughout at the time of origin or if it had produced massive lava flows from the interior at some time in its history," he said.[16]

The Eve of Apollo

On the eve of the first manned lunar landing in 1969, none of the controversies about the origin and evolution of the Moon had been resolved by the reconnaissance of Ranger, Surveyor, and Orbiter. Man now was ready to investigate in person. Although the program of nine lunar landings projected in 1969 would illuminate these questions, expectations of new discoveries from Apollo 11, first of the nine missions, were not high. The flight essentially was an all-up engineering flight test of the Apollo transportation system. This purpose was underscored by the character of the crew, all experienced test pilots who had flown in Gemini, the two-man space vehicle that preceded Apollo.

* One of several estimates of when these sightings first were recorded.

The Apollo 11 commander was Neil A. Armstrong, 38, a civilian test pilot, who had flown 78 combat missions in the Korean War as a Navy fighter pilot. After being graduated from Purdue University in 1955 with a Bachelor of Science degree in Aeronautical Engineering, Armstrong had flown the X-15 to 200,000 feet, the X-1 rocket airplane, and had tested most of the fighter aircraft in the United States inventory. On March 16, 1966, he flew the Gemini 8 mission with David R. Scott, the first to make rendezvous and dock with an Agena target rocket. A malfunctioning thruster on the Gemini forced the crew to disengage from the Agena and make an emergency landing. Quiet and efficient, Armstrong was selected as an astronaut in 1962. He was married to the former Janet Shearon of Evanston, Illinois. They had two sons, Eric, 12, and Mark, 6.

Armstrong's partner on the Moon was Air Force Lieutenant Colonel Edwin E. (Buzz) Aldrin, Jr., 39, a graduate of the U.S. Military Academy. Studious and introspective, Aldrin specialized in spacecraft guidance and orbital rendezvous. This was the subject of his doctoral dissertation at the Massachusetts Institute of Technology, where he received a degree of Doctor of Science in Astronautics. He, too, had been a fighter pilot in the Korean War. He was credited with destroying two MIG-15s and with damaging one.

Aldrin had been working closely with NASA as an Air Force aerospace officer when he was selected as an astronaut in 1963. He and his wife, Joan, had three children: James, 13; Janice R., 11; and Andrew J., 10. One of Aldrin's principal contributions to manned space flight was the demonstration on the flight of Gemini 12 with James A. Lovell in November 1966 that a trained man could work effectively outside an orbiting vehicle in free fall. He performed one of the most successful EVAs (extra-vehicular activities) in the Gemini program as a result of intensive preflight practice. On Apollo 11, he was the pilot of the lunar module Eagle.

Destined to remain in lunar orbit aboard the Apollo command module Columbia, while Armstrong and Aldrin walked on the Moon, was Air Force Lieutenant Colonel Michael Collins, 38. A graduate of the U.S. Military Academy, Collins served as an experimental flight test officer at Edwards Air Force Base, California, prior to being selected as an astronaut in 1963. He flew the Gemini 10 mission with John W. Young and performed three EVAs. They docked with an Agena target rocket that boosted the Gemini to a record altitude of 475 miles and then made a rendezvous with another Agena.

Although the scientific activities of Armstrong and Aldrin on the Moon were minimal compared with succeeeding flights, the scientific results of their mission were nevertheless historic. The samples proved that there was no life on the Moon, living or fossil, before man came there, and that the lunar soil would grow plants as well as or better than Earth soil. They set up scientific instruments that began to give investigators their first clues to the nature of the Moon . . . and to its complexity. From this viewpoint, the mission was of enormous value scientifically as well as technologically, but it was designed primarily as an engineering experiment.

During July 1969, as Apollo 11 stood on Pad A of Launch Complex 39, Kennedy Space Center, there was an aura of expectation that Apollo would make revelatory discoveries that would illuminate the secrets of Creation. Nearly a million people gathered on the beaches to watch mankind take the giant step. Russian trawlers hovered off the coast, sporting huge antennas. The world listened and waited.

Hardly ten centuries after venturing westward from Iceland, Western man was commencing the exploration of the Solar System.

Notes

1. Velikovsky, E., *Worlds in Collision*, Dell Publishing Co., New York, 1965; and *Pensée: Student Academic Freedom Forum*, Vol. 3, No. 1, Winter 1973.
2. Surveyor 5 Preliminary Report, NASA SP 164, December 1967.
3. *Science*, Vol. 160, June 7, 1968.
4. Surveyor 5 Preliminary Report, NASA SP 164, December 1967.
5. *Science*, Vol. 160, June 7, 1968.
6. Ibid.
7. Surveyor Project Final Report. Technical Report 32-1265, NASA JPL, June 15, 1968.
8. Ibid.
9. Ibid.
10. Ibid.
11. Rowan, L. C., "Orbiter Observations of the Lunar Surface," a paper presented at the AAS Symposium on the Physics of the Moon, December 29, 1966, AAAS Annual Meeting, Washington, D.C.
12. Urey, Harold, "The Contending Moons," *Astronautics & Aeronautics*, January 1969.
13. Ibid.
14. Kaula, William M., "Gravitational Field of the Moon," *Science*, Dec. 26, 1969.
15. Menzel, Donald H., "The Nature of the Surface of the Moon," *Physics of the Moon*, Vol. 14, AAS/AAAS Symposium, December 29, 1966.
16. Urey, Harold, *Science*, April 23, 1971.

4

Landing at Tranquility

At an altitude of 21,000 feet, the lunar module Eagle was descending toward the cratered surface of the Sea of Tranquility at a rate of 1,200 feet per second. Neil A. Armstrong and Edwin A. Aldrin, Jr., watched tensely as the radar altimeter and other displays on the LM console recorded the descent.

"You're GO for landing," announced Mission Control at Houston.

"Roger. Understand," said Armstrong. "Go for landing."

When the LM reached an altitude of 2000 feet, Armstrong and Aldrin were able to observe the moonscape in detail. They were about a mile uprange from the landing site, which they could identify roughly from the photos they had studied during training. Then they realized that the machine would touch down in or near the rocky rim of a crater about 200 yards in diameter, called West Crater. The area appeared too rough for a safe landing. They could see many big rocks, larger than 5 or 10 feet across.

"We elected to overfly this area in preference for the smoother spots a few hundred yards farther west," Armstrong related afterward. As they slanted down toward the west, they saw several relatively open areas between fields of boulders. They headed in on their final approach between two of these boulder fields toward a relatively smooth field of small circular and elongated craters. Feelers on the Eagle's landing pads touched the lunar soil, and the contact light flashed on the console.

"Houston, Tranquility Base here," said Armstrong. "The Eagle has landed." His words were barely audible above the uproar of cheering and applause in Mission Control where the landing was also recorded. It was 32 seconds past 4:17 P.M. Eastern Daylight Time on June 20, 1969.

"Roger," responded Houston. "Roger, Tranquility. We copy you on the ground. You've got a bunch of guys about to turn blue. We're breathing again. Thanks a lot!"

The landfall was later pinpointed at 0.67 degrees north and 23.49 degrees east about 130 meters north and 600 meters west of West Crater, just a bit to the right of the center of the Moon as you look at it from Earth. It was 20 kilometers south-southwest of the Crater Sabine D in the southwestern part of Mare Tranquillitatis. Surveyor 5 stood 25 kilometers north-northwest, and an impact crater formed by Ranger 8 was 68 kilometers to the northeast. Thus, the region was fairly well known. It had been photographed extensively not only by Ranger 8 and Surveyor 5 but also by the Lunar Orbiters and the two previous reconnaissance orbital missions of Apollos 8 and 10.

This region of the Sea of Tranquility is crossed by the northwest trending rays of splash-out from the impact that made Crater Theophilus, which is 320 kilometers to the southeast. About 15 kilometers west of the landing site, a north-northeast trending ray of splash-out has been identified, probably from Crater Alfraganus, 160 kilometers to the southwest, or from Tycho—or from all three providing a widespread sampling of lunar rock types. About 41.5 kilometers to the southwest was the promontory of the Kant Plateau, the nearest highland region.

The lunar module stood on ground consisting of unsorted, fragmental debris that ranged in size from particles too fine to be resolved by the unaided eye to blocks 0.8 meter across. This debris formed the surface layer that geologists called the regolith. It turned out to be quite porous and weakly coherent at the surface, and it graded downward into more densely packed material, as Surveyors had found.

The spacecraft was tilted toward the east 4.5 degrees from the vertical and yawed to the south 13 degrees. From the window of the LM, the crewmen saw a desertlike surface relatively free of rocks, with gentle rises and hollows pitted by craters ranging from a foot to 100 feet in diameter. The photographs they took of the region with the Hasselblad camera show the immediate area of the landing site to be deceptively smooth. To the crewmen the site appeared to be very rough. It had slopes, holes, and ridges, and the general topography consisted of undulations a meter or more in amplitude.

In the auditorium of the Manned Space Craft Center, Houston, a bespectacled engineer in a blue suit appeared on stage, his hands clasped before him like those of a minister about to deliver a sermon. He was Thomas O. Paine, administrator of NASA. He said that he had telephoned the White House and had reported the landing to President Richard M. Nixon. He said he had told the President: ". . . the Eagle has landed in the Sea of Tranquility, and our astronauts are safe and looking forward to starting the exploration of the Moon."

At 10:39 P.M., after both men had rested and checked their equipment, Armstrong opened the hatch, squeezed through it with his backpack containing his portable life-support system, and began climbing down the nine-step ladder. When he reached the second step from the bottom, he pulled a D-ring that opened a compartment on the outside of the LM and deployed a television camera. It began to photograph him as he reached the last step. There he paused and looked down. The LM footpads were depressed in the soil only an inch or two, Armstrong reported. "The surface appears to be very, very fine grained," he said. "As you get close to it, it's almost like a powder." Aldrin photographed the descent from inside the LM cabin with the Hasselblad. Armstrong's left foot touched the ground at 10:56 P.M. He then issued his oft-quoted landing statement, which came over the radio as: "That's one small step for man, one giant leap for mankind." Later, NASA amended the phraseology so that the utterance would be recorded in history as "one small step for *a* man." The article had been omitted either by Armstrong or in transmission. The astronaut was breathing hard; he was excited at being the Columbus of space and was not a man given much to rhetoric.

Armstrong stood for a moment near the ladder, getting his "Moon legs." Then he began to move cautiously, testing the feel of one-sixth gravity. "The surface is fine and powdery," he reported. "I can pick it up loosely with my toe. It does adhere in fine layers like powdered charcoal to the sole and sides of my boots. I only go in a small fraction of an inch, maybe an eighth of an inch, but I can see the footprints of my boots and the treads in the fine, sandy particles. There seems to be difficulty moving around as we suspected. It's even perhaps easier than the simulations. . . ."

From the LM window, Aldrin described the surface as "pretty much without color." Depending on which way you looked, the color ranged from a chalky gray away from the Sun to an ashen gray toward the Sun. Some rocks that seemed to have been fractured or disturbed by the LM rocket blast were coated with a light gray material on the outside but displayed a dark gray interior when broken. "It looks like it could be country basalt," Aldrin said.

Armstrong began to collect soil samples near the LM, using a scoop on the end of a pole so that he would not have to bend down in the bulky Moon suit. He scooped up two batches weighing a total of 1.4 kilograms (about 3 pounds) and stored the bag containing the sample in the left leg pocket of his space suit. This was a contingency collection, made immediately so that if some emergency required a quick departure before the main samples were collected the crew would not come back empty handed.

Meanwhile, he kept talking: "This is very interesting. It's a very soft surface, but here and there I run into a very hard surface, but it appears to be very cohesive material of the same sort. . . . This has a stark beauty all its own. It's like the high desert of the United States."

Armstrong then moved the television camera from its mount on the LM and set it up on a tripod 40 feet away, so that it could cover the area where the astronauts would set up three experiments and collect samples of rocks and soil.

It was time for Aldrin to climb down. Passing down the Hasselblad still camera to Armstrong, Aldrin backed down the ladder. Armstrong took pictures of him climbing carefully down. On the surface, Aldrin tested his stance and began the ungainly kangaroo-style walk. The two men photographed the stainless steel plaque on a leg of the landing craft that read: "Here men from the Planet Earth first set foot on the Moon July 1969. We came in peace for all mankind." The plaque listed the names of the astronauts and of President Nixon.

Aldrin opened a storage bin on the LM and took out a 12-inch tube in which a 4-foot length of aluminum foil was rolled around a telescoped aluminum pole and a folded crossbar. He extended the telescoped pole and pushed it into the soft soil, then erected the crossbar so that the aluminum foil would hang downward from it like some bright, ducal banner, facing sunward. The foil was designed to trap solar wind particles, which would be identified at the laboratory of Professor Johannes Geiss of the University of Berne, Switzerland, after the crew returned to Earth.

From the leg of the LM, the astronauts extracted a 3-by-5-foot nylon United States flag. It, too, was hung on a pole driven into the soil, but it was braced by a spring wire at the top to keep it extended on the airless Moon. When the flag was erected, both men stepped back and saluted it. President Nixon then called them on the Apollo communications network and congratulated them in what he described "has to be the most historic telephone call ever made."

Considering the magnitude of the event, the ceremonies were brief and perfunctory. For the first time men stood on another world, in full view of their own. The plaque and the flag might remain on the quiet, windless mare for millions of years and never be seen again, or it might become a shrine on a populated Moon.

The ceremony represented a new departure in the exploration of new lands. For the first time no national claim to territory was made. The plaque and the flag simply represented the people who came there first and the people who sent them. That, too, was a giant leap for mankind.

From another compartment of the LM, Aldrin extracted the passive seismic experiment package. It consisted of three long-period seismometers and one short-period, vertical seismometer for measuring both horizontal and vertical motions of the surface caused by moonquakes and meteorite impacts. These instruments, capable of recording the slightest tremor—even the footfalls of the astronauts—were set up about 30 feet from the LM. The system was powered during the lunar day by solar cells converting sunshine into electricity. At night when the temperature dropped to 279°F below zero, the system was warmed to 65° below zero by two radioisotope heaters 3 inches in diameter and 3 inches long. Each was fueled with 1.2 ounces of plutonium-238. This was the first nuclear device to be emplaced on the Moon.

Armstrong next set up a laser mirror (Lasser Ranging Retroreflector) designed to reflect a laser beam from an Earth telescope back to its source, the telescope. The mirror consisted of 100 high precision, fused silica corner reflectors set in an aluminum frame 46 centimeters (17.9 inches) square. In many respects the mirror was one of the most important experiments taken to the Moon in the Apollo program. By reflecting a ruby laser beam "fired" at it from the 120-inch telescope of the Lick Observatory, Mount Hamilton, California, near San Jose, or from the 107-inch telescope of the McDonald Observatory, Mount Locke, Texas, near Fort Davis, the mirror would enable scientists to determine more precisely than ever the distance between the Earth and the Moon. The mean distance had been calculated to 500 meters. With the laser ranging system made possible by the mirror on the Moon the error could be reduced to 75 meters. Aimed and fired through the telescopes, the laser beam made a splash of red light on the Moon about 2 miles in diameter encompassing the mirror. Some of the light was reflected back into the telescope lens by the mirror, making the round trip in 2.5 seconds.

In addition to refining the mean distance between Earth and Moon, the laser system of measurement made it possible for scientists to acquire new information on gravitation and relativity. It could reveal fluctuations in the Earth's rate of rotation as well as in motions of the polar axis. It could also indicate large-scale movements of the Earth's crust—relative, of course, to that little mirror on the Moon. For the first time scientists would be able to measure continental drift by sightings on the Moon. Of particular interest was the checking of a hypothesis that the Hawaiian Islands are drifting toward Japan at the rate of 10 centimeters (3.9 inches) a year.

The technique for pointing the beam had been confirmed during the Surveyor 7 mission by tests with an argon-ion beam that was seen when it illuminated the Surveyor 7 landing site on the Moon by Surveyor's television camera.

Armstrong achieved an accurate alignment of the mirror within one compass point east and west and leveled it to within 0.5 degree.

These three experiments—the solar wind composition test, the seismometers, and the laser mirror—took longer to set up and adjust than planned. The amount of time for collecting the rock and soil samples was thus reduced.

Working feverishly, both explorers picked up 16 kilograms of bulk samples in 14 minutes and spent another 5 minutes sealing the container. It took them 5 minutes and 50 seconds to get two core tube samples by driving a hollow tube into the ground near the solar wind experiment. Then only 3 minutes and 35 seconds were left on the time line for Armstrong to collect 20 "grab" samples weighing a total of 7 kilograms in the final moments of the EVA (extra-vehicular activity).

Most of the samples were undocumented as to their position when picked up. Their location and orientation to the surface had to be ascertained later, by Armstrong's panoramic photos of the surface before sample collection began and from photos taken from the lunar module window before the EVA started.

Back at Houston the Lunar Geological Experiment team was not happy about the haste with which the specimens were collected. They wanted documentation to determine the local geologic environment of each specimen and the relationship of each to other rocks and materials at the place where the sample was picked up. This information was essential in determining the processes that affected the specimen and its history.

Both Armstrong and Aldrin commented that the rocks they were seeing looked basaltic to them. At one point Armstrong said he noted vesicles in some rocks—tiny fissures made by the escape of gases during a period when the rock was molten. Later, however, he amended this observation and said perhaps the indentations he at first thought were vesicles actually were micrometeoroid pits.

With the 1.4-kilogram contingency sample, the 16-kilogram sample, and the 7-kilogram grab samples, the astronauts picked up 24.4 kilograms (53.68 pounds) of rock and dirt on the Moon, not counting the two core samples. On this first foray neither wandered very far from the LM. Both traversed a minimum total distance of 750 to 1000 meters. Armstrong made the most distant walk to a crater 33 meters in diameter east of the LM and photographed it near the end of the EVA.

During the Moonwalk surface temperature ranged from 0°F to 50°F as the Sun rose in the lunar east. Two days later the temperature had shot up to 180°F in the unshaded lunar desert.

At 12:54 A.M. on July 21, 1969, after 1 hour and 43 minutes on the surface, Aldrin climbed back up the ladder and reentered the LM. Armstrong followed at 1:09 A.M. He had spent 2 hours and 13 minutes on the surface. Nearly 12 hours later, at 1:54 P.M., the Moonwalkers ignited the LM ascent engine. It promptly boosted them off the lunar surface into a low orbit from which they expertly made rendezvous with their colleague, Michael Collins, the unsung hero of that expedition, who had been cruising in the orbiting command module while his buddies were making history on the Moon.

The first lunar traverse was over. And the men who made it were on the way home with the most costly specimens to be collected in the history of science or exploration.

Ellis Island

Because of the uncertainty about the existence of microorganisms on the Moon, the returning crewmen had to go through decontamination procedure and a quarantine lasting 21 days from the time of the liftoff from the Moon. After splashdown on July 24, 1969, they were required to put on biological isolation garments, which were passed to them through the hatch by a swimmer from the recovery ship U.S.S. *Hornet*, an aircraft carrier. The swimmer, lowered into the sea from a hovering helicopter, also was accoutered in one of the garments to prevent contamination by any lunar organism.

The crewmen then climbed out of the command module, attired in their biological isolation suits, and made their way into a life raft. The raft itself contained a decontaminant solution. The spacecraft's hatch was then closed to prevent any hypothetical organisms from escaping into the atmosphere. Aboard the *Hornet* the first men on the Moon, looking like visitors from another planet in their quarantine garb, were hustled into a trailer, which was hermetically sealed. There they stayed until they were flown back to the Manned Spacecraft Center, Houston, where they were dumped into a quarantine section of the Lunar Receiving Laboratory.

No sign of any lunar organism was ever found in exhaustive tests of Moon rocks, Moon soil, and, of course, the Moon men. After the Apollo 14 mission in which the decontamination and quarantine procedures were relaxed the whole process was discarded. The evidence for life on the Moon was negative.

Moonquake

As the Apollo 11 crew headed back to Earth, there came a signal from the passive seismic experiment back at Tranquility Base. It was logged at 12:20 P.M., Central Daylight Time, July 22, 1969. It was a seismogram of what appeared to be either a Moonquake or a meteorite impact at some distance away from the seismometers.

Could the impact have been the crash of Luna 15? This was the vehicle the Russians sent to the Moon at the same time as Apollo 11, presumably to pick up a small sample and return it to Earth. Luna 15 crashed in the Mare Crisium the same day, so that this attempt to upstage Apollo 11 with a robot Moonscoop failed.

At the Manned Spacecraft Center the seismic experiment investigators were certain that whatever made the seismogram was not Luna 15. That vehicle was much too small and its impact too distant to be recorded, even though the sensitivity of the seismometers at Tranquility was very high because of the low background noise on the Moon. Compared to the Earth, the Moon was a quiet place, the investigators began to realize.

The seismic experiment investigators had been watching the signals from the apparatus since it had started recording while the astronauts were still on the Moon. It had picked up their footfalls, their movements climbing the ladder, the impacts of their Moonshoes, and other gear they threw over the side to lighten the LM just before takeoff and, of course, the liftoff of the LM. With such a sensitive instrument, Armstrong had remarked drily, it would be pretty hard for anybody to get away with anything on the Moon.

At first the seismic event signalled on July 22 was interpreted by the investigators as a probable Moonquake centered far away. Members of the seismic team consisting of Gary Latham and Maurice Ewing of the Lamont-Doherty Geological Observatory, Frank Press of the Massachusetts Institute of Technology, and George Sutton of the University of Hawaii considered it likely that what they had seen was the first Moonquake ever recorded.

The seismic waves were recorded in a sequence that indicated the presence of terrestrial-style crust. The recording began with high frequencies first and then displayed the lower ones. "If we were to see that on Earth," said Press, "we should say that it corresponds to an earthquake some thousands of miles away and that the dispersion—the variation in the length of the wave with time—was due to the presence of the crust of the Earth—that would be the interpretation we would make knowing that that event occurred on the Earth. Now, so far as this event occurring on the Moon is concerned, I would guess off-hand from our experience that this is an event that occurred on the order of a few thousand kilometers away—one or two thousand—and that it is the first evidence for layering in the Moon that we have."[1]

Press said that if the seismogram was caused by a meteorite impact, the meteorite would have been large—a relatively rare event on the Moon. He tended to feel it was actually a Moonquake. It was analogous to a California quake with a magnitude of 4 to 5 on the Richter scale recorded on the East Coast.

"In fact," he added, "I'm going to claim a case of champagne from a colleague in California on a wager as to whether or not there are Moonquakes on the basis of that one event."

Ewing agreed. "I think there are no doubts at all that the dispersion shown in that seismogram indicates that the signal is a surface quake from either a Moonquake or a large impact."[2]

The arrival of the high-frequency or short-period waves ahead of the longer-period or lower-frequency waves suggested the layering Press referred to. It meant that the short-period waves were traveling faster than the long-period ones and this indicated a two-layered medium, such as a crust overlying a mantle, as on Earth. On the Moon, however, this single event might also reflect another kind of crust if one existed— a much thicker one than the Earth's crust.

It was clear that the team believed that the event was evidence for a layered Moon, with at least a crust and a mantle, if not a core.

Ewing commented: "If the inference about a crust turns out to be substantiated by further indications of a similar trend, then we will feel quite certain that at one time in its history the Moon has been sufficiently hot to allow the matter of which it's composed to differentiate into a lighter crust and a heavier mantle."[3]

But no other event was recorded by the Apollo seismometers. The package was heating up under the rising sun, and the heat was threatening the continued life of the radio transmitter that relayed the seismic signals to Earth receiving stations. After 21 days of operation, the transmitter and receiver components of the package failed and the experiment stopped.

Historically, the seismic event recorded on July 22 was the first evidence of a differentiated Moon and, at the time, was an exciting piece of data. But it did not survive in the Preliminary Science Report that was published at the end of October 1969. On further study and consultation, the signal that the team hailed so enthusiastically at first receded into ambiguity. The discovery of July 22 was erased after additional data indicated that these signals came from the instrument itself, not from events in or on the Moon. According to Latham, it was not until the seismic team had recorded several man-made impacts that the presence of a crust on the Moon was confirmed.[4]

Record of the Rocks

Preliminary examination of a portion of the rocks and fines (fine soil) returned by the crew of Apollo 11 to the Lunar Receiving Laboratory at Houston on July 26, 1969 began immediately and continued for two months. Samples were manipulated through glove boxes in which they were sealed to avoid any possible contamination of the terrestrial biosphere by the lunar lithosphere. A glove box is a sealed compartment in which long-sleeved rubber, rubberoid, or plastic gloves are affixed to one transparent wall of glass or clear plastic. By putting hands and arms into the gloves, the observer can manipulate the sample without breaking the seal and exposing the sample to the atmosphere.

When the preliminary examination of the lunar rocks and fines was completed, members of the Preliminary Examination Team made their report public at a press conference on September 15, 1969, at the Museum of Natural History of the Smithsonian Institution in Washington.

At the Apollo 11 landing site, the investigators said, the lunar surface consisted of unsorted fragmental debris (the regolith), the bulk of which was made up of fine particles. Rocks were embedded in the regolith matrix, and evidence showing they had been shifted or turned suggested that they might have been splashed out of craters by the impact of incoming objects from space.

The investigators said they were fascinated by an unexplained erosion process, first observed in the Ranger, Orbiter, and Surveyor photos and now confirmed by Apollo 11. Erosion on the Moon is unlike any process observed on Earth, the team reported. It apparently was the production of impacts by space dust and of nuclear particle bombardment by the solar wind.

The rocks at Tranquility Base were igneous rocks. They were formed by cooling or crystallizing from a melt—a lava that had been formed by the heating up of earlier rocks. What the source of heat was could be surmised. It was radioactivity, which is also the source of the Earth's heat.

The team was looking at second and possibly third generation rocks. These, as the alpha-particle scattering analyzer of Turkevich and his associates had found on Surveyor 5, were generally similar to basaltic rocks on Earth but also significantly different. Chemically, the differences were marked by unusually high concentrations of titanium, zirconium, and yttrium. These are refractory metals (metals that do not melt quickly) and hence would accumulate in a melt as more volatile elements boiled off.

Chemical and mineralogical differences between the lunar basalts and those known on Earth implied that either the parent rocks, which melted and then cooled to form the Tranquility rocks, were different than terrestrial rocks or the processes of melting and recrystallization were different. The nature of the difference and how it came about were to become key questions later on as the origin and history of the Moon became increasingly involved in controversy.

Chronologically, the rocks brought back from Tranquility astonished the investigators by their relatively great age. They were found by measurements of radioactive decay to be 3.5 to 3.7 billion years old. Just to pick up rocks of that age on a first visit was exciting, because geologists had searched the Earth for centuries before finding rocks that old in West Greenland. The indication that common rocks on the Moon were as old as the most ancient ones ever found on Earth encouraged the lunar investigators to believe they would find lunar rocks as old as the Earth itself, perhaps forming the primeval crust of the Moon, as the Apollo expeditions continued and the range of astronaut extra-vehicular activity on the lunar surface increased with experience and confidence.

Rocks were divided into four groups. Type A consisted of fine-grained, vesicular, crystalline rocks that some geologists called ferro-basalts. These were distinguished by vesicles or tiny, hollow tubes through which gases escaped when the rocks were molten. Type B was a medium-grained, vuggy, crystalline rock called microgabbros. A "vug" is a small cavity also created by outgassing. Type C was a breccia, which is a conglomeration of fragments of different rock types, minerals, and glass cemented together by heat and pressure. Type D was dust—the fines—forming the mass of the soil.

There was no question that the rocks had cooled from lava or near surface melts. Whether the heat that produced the melt came from volcanic sources or from the impact of meteorite bombardment was debatable at the time of the preliminary examination report. Sizes of the mineral grains and of the vesicles as well as the shapes of the rocks suggested that they formed near the top or near the bottom of a lava flow or a lake of lava. The absence of hydrous mineral phases was notable. These rocks were formed without water, an observation confirming the pre-Apollo 11 suspicion that there was no water on the Moon. At least there had been no water when these rocks were made.

The most striking features of the chemical composition of the Tranquility rocks were the high concentrations of titanium, zirconium, and yttrium, compared with Earth rocks and with meteorites. Analyses showed up to 12 percent titanium oxide, a concentration rare in Earth rocks. Whether these concentrations were a local or general feature in the maria remained to be seen. They indicated, however, that the rocks had been extensively "cooked" under very hot conditions, probably near the surface. This led to an early conclusion that the upper parts of the Moon had been molten between 3.5 and 3.7 billion years ago. Compared to chondritic meteorites, the rocks from Tranquility were quite low in nickel and cobalt. No nickel at all was detected in several samples. The pre-Apollo theory that the Moon was meteoritic in composition could be rejected. It was far from that. Nor was it terrestrial. These rocks did not come from any source resembling the mantle of the Earth. Geochemically, the results of the first assay of the lunar rocks weakened the theory that the Moon had been torn from the Earth. The concentrations of uranium and thorium were near the values for Earth basalts. The gross mineral and chemical compositions were similar— but there were important differences.

This unique composition of the rocks returned from Tranquility was evidence that the primeval material from which the melt that formed these rocks was derived was chemically different from the Earth's mantle. It suggested that the Moon was indeed a separate planet, as Urey hoped it would turn out to be.

Minerals in the lunar samples were all known on Earth. They included a type of feldspar called plagioclase and assemblies of pyroxene and olivine. A major constituent of the fine-grained igneous rocks was ilmenite, an iron-titanium oxide. The fines consisted mainly of dust particles ranging from 30 microns to less than 1 millimeter in size. About half of the fines consisted of glass and tiny, shiny spheroids, suggesting wholesale impact metamorphism from meteorites.

Preliminary investigation by mass spectrometry failed to find any organic matter in the rocks or fines. No evidence of life, living or fossil, was detected. No evidence of pathogenic organisms, bacteria or viruses, was found in germ-free mice, fish, quail, shrimp, oysters, tissue cultures, insects, plants, and even one-celled paramecia exposed to lunar soil. Until the astronauts came Tranquility had no life. It was as sterile as the beginning of time. Or so it seemed in the preliminary examination.

In the coarse breccia, relatively large amounts of the noble gases— helium, neon, argon, krypton, and xenon—were found, implanted there by the solar wind.

Erosion was obvious. Most of the rocks were rounded where their edges were exposed. Some appeared to have been sandblasted, possibly by the infall of space dust. All rocks showed glass-lined surface pits, apparently from the high-energy impact of small particles. The discovery that the interior of the breccia samples contained noble gases derived from the solar wind implied that they were formed at the surface from fragments previously exposed to the solar wind. However, the mechanism of their formation was not clear. What had happened to cement these bits and pieces of pulverized rock together to form the coarse conglomerate?

There were more questions about the rocks after the preliminary examination than had been thought of before. A photograph taken by Armstrong of one basalt rock with a sharp point on one side and a rounded surface on the other showed that before the astronaut picked up the rock the sharp point was up and the rounded part was buried in the soil. The rock had been turned over, exposing the noneroded point. Not all the rocks Armstrong and Aldrin recovered were photographed *in situ*, but this one was. Had it not been, geologists looking at it on Earth would have surmised that the rounded part was up because it appeared to have been eroded by the solar wind or space dust. The fact that the sharp end was up could mean only one thing—that the rock had been tumbled. Other photographs of samples showed rocks with eroded faces down. They suggested that all the rocks had been thrown about, probably not once, but many times. What agency was throwing rocks around on the Moon? One answer seemed to fit—meteorite impacts.

Evidence that the surface was not as motionless as imagined was surprising indeed. The accumulation of argon-40 from the decay of potassium in eight of the Tranquility rocks gave a minimum age of 3.5 billion years. This, too, was surprising. For it could be assumed that the rocks of the maria were younger than those over which the maria had formed, so that rocks as old as the oldest ever found on Earth were actually the younger rocks of the Moon.

But how long had these rocks been lying on the surface at Tranquility? This could be determined roughly by determining the accumulation in the rocks of such isotopes as helium-3, argon-38, and neon-21. These are stable isotopes that are created by the bombardment of cosmic rays. Thus, the length of time the rocks had been exposed to cosmic rays could be ascertained. It would tell how long they had been on or within a meter of the surface. The eight rocks that had been potassium–argon dated yielded cosmic ray exposure ages of 20 to 200 million years. Yes, there was movement on the Moon, but it was very slow, and at Tranquility it had occurred a long time ago. Some of the eight rocks had been lying at Tranquility since the Triassic Period of geologic time on Earth when dinosaurs roamed the land. The Moonscape of Tranquility was incredibly ancient by Earth standards. Although there had been motion in the far past, it had been still at Tranquility since before the time of Man.

The Hot Moon

The preliminary radioactive measurements of rocks and fines showing that the concentration of uranium and thorium was about as high as that in terrestrial basalts led to an important hypothesis. If the amounts of uranium and thorium in the Tranquility samples were distributed uniformly throughout the Moon, the planet would presently be quite hot and would have active volcanoes, like those on Earth. But the Moon now appeared cold. Taking the age of the basaltic rocks into account, it seemed logical to assume that radioactive heating had melted a portion of the upper Moon some 3 to 4 billion years ago. But whether the Moon might still be hot at depth remained debatable.

On the surface, differences between the rocks and fines began to appear—differences hard to explain. On Earth, dust and soil are assumed to be the products of the disintegration or erosion of rocks and therefore should have the same basic history. But indications were appearing in the dating experiments that in some instances the lunar soil was older than the rocks embedded in it. This was a puzzling circumstance suggesting that the rocks had come from somewhere else— possibly they had been splashed out of distant craters by meteorite impact. Another difference appeared between the breccia and the igneous rocks. The content of noble gases in the breccia was a thousand times higher than that in the basaltic rocks. In this respect, the breccia were closer to the topsoil that also contained up to three orders of magnitude more noble gases than the basalt.

There was another important difference. On Earth the most interesting rocks from a magnetic point of view are the igneous rocks, because they indicate the intensity and direction of the geomagnetic field at the time they crystallized. On the Moon, the igneous basalts were magnetically inert, but that did not mean that magnetism was absent. Astonishingly, the composite rocks, the breccias, contained a detectable amount of remanent magnetism. The discovery of magnetism in these rocks was another surprise, for most investigators had assumed that the Moon always had been without a magnetic field. But what had magnetized the breccia?

"We don't know what the magnetic story is," said the late Paul Gast, then a geophysicist at Columbia University, "but we do know that there are some very strange magnetic properties in these rocks which were not expected."[5]

As I have mentioned earlier, both the similarities and differences in lunar and terrestrial basalts were striking. Lunar basalts were higher in titanium and chromium and lower in aluminum than terrestrial ones. This meant that the sources of the basalts on each planet—the sources of the melt from which the rocks crystallized—were different in composition. This difference implied that the Earth and Moon accreted separately, under different conditions of temperature in the solar nebula, and probably at different distances from the Sun. If one accepted this reasoning, one then could dismiss the fission theory that the Moon had broken off the Earth.

Differing from both the fines and the breccia, the basaltic rocks were unique in another way. They contained more radioactive elements and were denser than the whole Moon.

These findings, made public in Washington hardly two months after the landing of Apollo 11, established a new science of lunar geology. The definition of geology itself was expanded to include the study of all the terrestrial planets.

The task for the next expedition to the Moon aboard Apollo 12 was clear. The lunar explorers would have to document their samples more carefully than the Apollo 11 crew had time to do. There simply was not enough time for Armstrong and Aldrin to take photographic reference and stereo pictures of the samples *in situ* so that the sites from which each sample came, and its orientation, could be identified in the photos back at the Lunar Receiving Laboratory. This was particularly unfortunate in the case of the breccias. When it was discovered that they had remanent magnetism, it was important to determine their orientation on the surface and their spatial relationship to each other. But this was not possible because of lack of photographic records. For Apollo 12, Charles Conrad, Jr. and Alan L. Bean, the two astronauts who would land on the lunar surface, were being trained to do a full job of field geology.

The Glazed Surface

One of the curious discoveries by Armstrong and Aldrin at Tranquility Base was a scattering of glassy patches on the top of clumps of soil lying at the bottom of shallow craters. The glassy patches ranged in size from about 0.5 millimeter to about 1 centimeter. They were concentrated on the top surfaces of the clumps and on points and edges. Droplets of hot liquid seemed to have run down an inclined surface for several millimeters like hot icing on a cake and then congealed.

This strange glazing showed up in eight photographs that Armstrong took with a special camera developed by the Eastman Kodak Company. The camera was mounted on the lower part of a walking stick so that the astronaut could take a magnified picture of the surface without bending over in the bulky Moon suit. Armstrong could take the close-up photo by squeezing the handle of the stick that fired a flash, tripped the shutter, and advanced the film. Focus was established by touching the bottom of the stick to the ground.

Each of the photos covered an area 7.6 centimeters square and revealed details as small as 80 micrometers. The glazing appeared in photos of clumps at the bottom of small craters 6 inches to 2 feet in diameter. Armstrong had put the stick camera down at these places in order to photograph the structure of the small craters. Although he had noted the brightness of the glaze, he was surprised later when he saw in the developed photographs the glassy droplets and runs clinging like frosty dew to tiny pinnacles and stems of the fragile soil. The glaze appeared only in small craters, not in large ones.

The glazing phenomenon attracted the attention of the Cornell University astronomer Thomas Gold, who noticed immediately that some glazes were oriented in the same direction. This suggested to him that the glaze had been formed on the top of the clumps by an intense burst of radiation heating, and that its origin could only have been the Sun.[6] The glazing was different from the shock-produced glassy beads commonly found at the Tranquility site. It could not have been the result of the impact that made the craters because the frail, powdery structures on which the glaze appeared would not have survived.

The finding of the glaze only in the small craters enhanced the solar origin theory, for the noontime temperatures at the bottom of such craters could be expected to be about 11 percent higher than the temperature at the surface. Gold reasoned: If the average noontime temperature on the equator is 394° Kelvin (K), the temperature in one of the small craters would be 437°K.* But it would take a temperature of 1400°K to produce the glaze on the basalt.

If the excess heating came from the Sun, it would require that the Sun's luminosity increase more than 100 times its present value for a

---

* Degrees Kelvin (developed by Lord Kelvin) is a scale of temperature based on absolute zero, a condition when molecules stop moving. The K reading is always 273 degrees higher than the Centigrade reading for the same temperature. Thus, 394°K is 121°C or 249.8°F.

period of 10 to 100 seconds, Gold calculated. Such an outburst far exceeds any known solar flares. It would be closer to the phenomenon of a nova—a sudden, intense flare-up in another type of star.

Gold based his estimate of the flare's duration of 10 to 100 seconds on the known rate of heat transfer in the lunar material. If the heating had lasted longer than this range, the area of glaze would be larger; if it had lasted less than 10 seconds, the glaze would have been smaller or even microscopic.

Moreover, he reasoned, the flash of heat must have been recent enough so that the infall of micrometeorites would not destroy the clump structure. It seemed unlikely to him that the glazed clumps could be older than 100,000 years and probably had not endured more than 30,000 years.

From all of this reasoning, Gold deduced that the Sun had put out a minor, novalike outburst of energy between 30,000 to 100,000 years ago. The effects of such a heat discharge on the surface of the Earth would be dissipated by the atmosphere, but the flash could have had serious effects on the upper atmospheres of the Earth, Venus, and Mars. It could have swept away part of the Earth's helium outer atmosphere and all of the atmosphere of Mars, leaving only the small amount replenished from the polar caps.

Flash heating of the surface of Mercury might have glazed an entire hemisphere. Speculating further, Gold mused that the solar heat flash could have been caused by a large asteroid or comet falling into the Sun. The energy of impact would be converted into radiant heat. This possibility was suggested to Gold by a British astronomer, Fred Hoyle.

Although the idea that the Moon might yield up clues to past history of the Sun was attractive, a number of lunar scientists found Gold's interpretation of the origin of the glazing much too speculative to take seriously. Scientists at the Manned Spacecraft Center suggested that the droplets of glaze could be explained more credibly as the splash-out of molten rock heated by the impact of an incoming projectile. Gold hoped that the Apollo 12 astronauts might find more evidence for his theory, which would require that the glaze be Moon-wide.

The Cosmogenic Conundrum

Other "leads" to physical processes in the Solar System and to its origin flowed out of the Apollo 11 findings, which produced more cosmogenic speculation than any other scientific event in modern times. None of the four theories of the Moon's origin could be proved or disproved by the new data. There was enough to support the conflicting theories that the Moon originated as a twin of the Earth, that it was captured by the Earth after originating somewhere else, that it fissioned from the Earth or with the Earth from a super planet, and that it coagulated from debris in the Earth's orbit.

For John O'Keefe at the Goddard Space Flight Center, the Apollo 11 results supported the theory that the Moon fissioned from the Earth shortly after the Earth's core was formed. That was the reason nickel was so scarce on the Moon—it had gone to the Earth's core, he said.[7]

A fission model for the Earth–Moon system is not a new idea. As I have related earlier, it was proposed by Sir George Darwin early in the century. O'Keefe and others (notably, the British astrophysicist R. A. Lyttleton) went a long step further. They suggested that the Earth and the Moon were fissioned from a parent body.

O'Keefe theorized that such a protoplanet would have a deep atmosphere of light gases surrounding a shell of heavier silicates. It would have a metallic core formed by the center-seeking migration of liquid short-lived radioactive isotopes or because of the heat of accretion— material falling into it. O'Keefe thought that the heading made the light gases boil off. The loss of mass increased the rate of the planet's spin. As rotation accelerated, the mass of the protoplanet became gravitationally unstable. The rate of spin exceeded the adhesive force of gravity to keep the masses together and a piece, or pieces, of the protoplanet split off, whizzing off into the interplanetary medium like a shot from a sling, ultimately to go into orbit around the larger mass.

Cosmic Uniformitarianism

The dynamics of the break-up have been studied mathematically. They show that the larger pieces have a mass ratio of 10 to 1 (the larger is 10 times the mass of the smaller). There are droplets with 1 percent or less of the mass of the parent that form between the larger pieces, according to Lyttleton.[8] These or some of them may fall back into the major pieces or become satellites of the larger one.

In the case of the Earth–Moon fission, the mass ratio was not 10 to 1 but nearly 100 to 1. O'Keefe explained: The parent and daughter were revolving about each other at first, with a period of 4 hours, each rotating on its axis in about 2 hours. Huge tides were raised on each body, heating both to high temperatures. The Moon became gaseous and lost about 90 percent of its mass. In this way, O'Keefe sought to account for the Moon's loss of volatile elements, especially hydrogen, and consequent loss of water, plus the increase in refractory or heavy elements in the residue.

The fission school of the origin of the Moon—and, indeed, of the origin of the Solar System—has many ingenious ways of explaining observed phenomena. But like the accretion theory of Urey and others, wherein big planets accrete out of little ones, like a snowball rolling downhill, it does not explain how the snowball starts in the first place in a solar nebula of dust and gas. The genesis of planets, and of stars, remains unaccounted for. Can it be found on the Moon? Urey had said that the terrestrial planets were formed by the aggregation of the Moon-sized "planetesimals." The Moon was one that was left over from this process. But how did the planetesimals start accreting? A number of theoretical explanations have been proposed, but none has become generally accepted at this writing.

Fundamentally, however, both accretionists and fissionists sought a uniformitarian answer to the problem of the origin of the Solar System. In the eighteenth and nineteenth centuries, as was noted earlier, uniformitarianism in geology meant simply that those processes that built the Earth could be seen at work today and that the Earth we know, with its mountains, rivers, and seas, is not the product of accidental catastrophe, but of observable and predictable evolutionary processes. These processes could be identified by observing what was happening now.

In the Solar System the processes of both accretion and break-up are observable. The Earth accretes a thin mantle of cosmic dust weighing perhaps 100 million tons a year. The Moon similarly is accreting as dust, meteorites, and solar wind particles impact it. At the same time, the existence of meteorites, asteroids, and comets shows us that break-up has occurred. Does there exist a balance between growth and disintegration?

Both O'Keefe and Lyttleton suggested that accretion continued up to a point where break-up occurred as a result of spin-up. In the cosmogony of Lyttleton, primitive objects condensed in the solar nebula of dust and gas to a huge size. The nebula itself is rotating. It has angular momentum or vorticity of its own. Where this comes from need not be discussed here. It is enough to say vorticity is there.

Lyttleton surmised that the condensing body acquires the angular momentum (rotation) of the nebular vortex. As the body snowballs, gathering more and more material from the cloud of dust and gas in which it somehow started to accrete, the angular momentum of the body per unit of mass increases. This is the obverse of O'Keefe's idea that angular momentum increases because radioactive heating boils off the light gases.

In any event, when the angular momentum per unit of mass exceeds the force of gravity required to hold the mass together, this enormous primeval body splits. The two main pieces, a big one 10 times the size of the small one, separate and droplets form between them.

Lyttleton suggested that this mechanism could have produced the terrestrial planets, including the Moon.[9] Mercury, Venus, Earth, the Moon, and the large satellites of Jupiter could have been formed as droplets from the break-up of a massive protoplanet. Jupiter would be the larger of the two main pieces. The smaller piece is no longer around. There is no body in the Solar System with one-tenth the mass of Jupiter. All of the terrestrial planets and moons combined have hardly one-hundredth of Jupiter's mass, so they could be considered "droplets" relative to the giant protoplanet.

What happened to Jupiter's junior partner after the break-up? Lyttleton got rid of it by having it fly out of the Solar System entirely and disappear into interstellar space. This is not so far-fetched if his explanation is considered.

The speed necessary for an object to escape from the surface of Jupiter or any of the other outer planets is considerably higher than their orbital speed around the Sun. In order to escape the Jovian surface, an object would have to be accelerated to 32.9 nautical miles per second, more than 5 times Earth escape velocity; Jupiter's orbital speed around the Sun is only 7.04 nautical miles per second, less than half of Earth's. This means that any object whirled away from a protoplanet in the outer marches of the Solar System might well have sufficient velocity to escape from the Solar System. Obviously, the droplets forming the terrestrial planets and moons did not achieve solar escape velocity, or lost it, inasmuch as at least 12 of them, including the four Galilean satellites,* went into orbit around Jupiter and the others were decelerated, somehow, to fall into lower orbits around the Sun.

Terrestrial Superplanet

An alternative case is considered by Lyttleton to account for the Moon. If terrestrial planets formed within the nebula independently of the low-density outer planets, it is possible that the Earth and Mars are pieces of a superplanet and the Moon is a droplet from the break-up. The Earth–Mars mass ratio is 9 to 1 and the Moon is less than one-hundredth of their combined mass. In this case, the moons of Mars, Deimos and Phobos, must be considered captured asteroids, an idea that is supported visually by the photos from Mariner 9.

Why wouldn't Mars have been hurled out of the Solar System, as

---

* So called because Galileo saw them in his 30-power telescope.

Jupiter's little brother was supposed to have been? Lyttleton points out that the velocity of escape from the inner planets is much lower than their orbital speeds around the Sun. To escape from the surface of the Earth, a rocket must be accelerated to 6.04 nautical miles per second; the orbital speed of the Earth around the Sun is 16.06 nautical miles per second. Thus, the piece called Mars could have escaped from the Earth by the acceleration of the break-up, but its velocity would not have been sufficient to enable it to escape from the Sun. It might simply have gone into a higher orbit around the Sun—which, in fact, is where it is now.

One piece of evidence supporting this idea is the fact that the rotational speeds of Earth and Mars are both close to 24 hours, which could have been the rotational speed of the superplanet. Lyttleton suggests that Venus and Mercury might be another pair of pieces from another superplanet.

Support for fission could be found in an analysis of the size and distribution of lunar craters. Zdenek Kopal of the University of Manchester, England, on leave at the Lunar Science Institute, Houston, in 1971, concluded from a study of craters that the bodies that made them could not have been in solar orbit. Because the angles of impact were distributed at random, these projectiles must have been in orbit around the Earth–Moon system, he stated.[10] Kopal speculated they were left over from the process that led to the formation of the Earth and Moon, which eventually swept them up. This would imply that the Earth and the Moon were gravitational partners during the first few hundred million years of their existence, Kopal said, but it did not rule out capture. However, if the Moon was captured, that event took place before the major cratering process started.

The question of the Moon's origin was probably the most popular subject of speculation arising out of the examination of Apollo 11 rocks. Next came the evolution of lunar features, particularly the maria. At the University of Chicago, a petrologic model of the Moon was proposed early in 1970 by geologist Joseph V. Smith and his associates in the Department of Geophysical Sciences.[11] The son of an English farmer, Smith had the profound belief in direct observation that had characterized the work of Hutton and Lyell two centuries earlier. In the fine-grained ferrobasalts and in the medium-grained, vuggy microgabbros from the Sea of Tranquility, the geologist perceived the products of extensive melting near the Moon's surface. The story in the chemistry of the rocks was all quite clear and straightforward. Differentiation of the melt and crystallization were so extensive that a large part of the lunar interior must have been molten at some time in its development.

This led Smith and his colleagues to visualize a model for a molten

Moon, convecting, fractionating, and producing shells of olivine and pyroxene cumulates to form an anorthosite crust. The underlying hot magmas, rich in titanium, were produced at a late stage of melting. They formed a layer just beneath the crust. When a meteorite ripped through the crust, the magma flowed out to form portions of the maria.

Because of the Earth's tidal attraction, the magma became concentrated on the near side of the Moon, the group concluded. This explained the strange lack of maria features on the far side.

The ferrobasalts and microgabbros at the Sea of Tranquility recrystallized from magmas with a composition somewhat similar to that of terrestrial basalts. The lunar basalts, however, were depleted in volatile elements and enriched in refractory ones, particularly in titanium and iron. If their low volatile content was a general feature of the Moon, it suggested widespread, perhaps Moon-wide melting in the past, when the volatile elements boiled off.

The early melting of the Moon was accompanied by convection that transferred most of the heat-producing, radioactive material to the surface. The Moon then cooled from the inside outward. It was possible that a small metal core formed, chiefly of iron. If so, it would have drawn the nickel from the silicates of the mantle and crust—that is, if the whole Moon was as low in nickel as the Apollo 11 rocks.

With this model, the Chicago group sought to reconcile the chemistry of the rocks, which clearly indicated early cooking, with the evidence that the Moon was no longer hot—at least near the surface. If the Chicago version of what happened to the nickel was true, it would knock out one prop from O'Keefe's fission theory. One question remained, however, in this synthesis. What was the nature of the original material that had melted? What were the parent rocks made of? The Chicago geophysicists had an answer for that, too. Fragments of anorthosite found at Tranquility, apparently splashed into the mare by meteorite impacts from the highlands, probably represented the original crust of the Moon, much of it demolished by the infall of planetoids that punctured the crust, squeezing the magma out like the juice of a perforated grape.

This model challenges the predominant conception of a cold Moon persuasively set forth by Gilbert in 1892 and by Urey 60 years later. Gilbert and Urey accounted for surface melting as the result of heat from meteorite impacts. Gilbert, and later Ernst J. Opik (1967), had suggested that the craters were formed by infalling bodies of low velocity and that most of these bodies had been stored in the Earth's orbit as a residual ring of sediment from which the Moon itself had accreted.

The sediment-ring hypothesis was later reviewed by Ringwood on the basis of data from Apollo 11 and the expeditions following it.

The Chicago group was not without historic support on the hot Moon front. Baldwin in 1949 had argued that internal heat produced volcanism of a large scale. The late Gerard Kuiper who was one of the world's leading authorities on the Solar System and a principal investigator in Project Ranger, had asserted in 1954 that the Moon was nearly melted throughout by its own radioactivity. The melting episode, he calculated, took place 500 million to 1 billion years after its formation.[12] Kuiper thus accounted for the formation of the maria during this early melting epoch.[13]

Theories that the Moon had been hot enough to have melted partially or wholly ran up against the curious fact that its center of mass was 2 kilometers offset from where it should be if the planet had been in hydrostatic equilibrium. Such equilibrium would be expected if the Moon had ever been melted throughout.

The deviation of the Moon's center of mass was a cold Moon argument. It suggested that the Moon had been rigid enough to prevent equilibrium from asserting itself and distributing the lunar mass so that its center would match the center of the spherical figure. If the Moon had been heated all the way through, the deviation would have disappeared.

This reasoning did explain the discovery that the rocks found at Tranquility had been cooked a billion or more years after the Moon was formed. It was possible that the cooking had taken place near the surface and that the center was cooler than the radioactively hot regions nearer the surface.

The Chicago group expressed the belief that an early melting of the Moon with thorough convection, transferring the bulk of the heat producing, radioactive elements to the surface, offered the best opportunity of reconciling an early hot Moon with a cooler present one.

Cold Moon supporters could explain the recrystallization of the surface rocks from a melt by showing how it was possible for surface lava to have been produced by the heat released by the collision of a large meteorite or asteroid. One of the proponents of the hot Moon idea was Paul Gast. He said it must be assumed that most of the igneous rocks on the surface were the product of partial melting in the interior. This was followed, he said, by upward transport of an igneous liquid. But it did not preclude the possibility, he added, that many rocks of the surface had undergone a second period of melting due to collisions.[14]

One of the arguments against a hot Moon was the apparent absence of a magnetic field. Because it is widely believed to be generated by the rotation of a body with a metallic core, like the Earth, the lack of such a field suggested that no core was present in the Moon. The apparent absence of an iron core was also supported by the Moon's overall density, which was about the same as that of the Earth's mantle but lower than that of the entire Earth when the Earth's core was averaged in.

But in the ongoing analysis of the Apollo 11 rocks, the detection of a remanent magnetic field in some rocks strongly suggested that a lunar magnetic field had existed in the past. Thus, the Moon may have had a magnetic field, not one as powerful as the Earth's, but one that was strong enough to leave a magnetic trace in the lunar crust.

The theory of planetary magnetics required two elements: an iron core and a rate of rotation rapid enough to produce magnetic force. The Moon now rotates on its axis very slowly compared with the Earth, and its period of rotation (27.322 days) might also account for the lack of a magnetic field with or without a metallic core. Evidence that a magnetic field had once existed suggested that the Moon may indeed have a core and that its period of rotation was formerly much faster. This idea seemed to be consistent with the possibility that tidal stresses exerted by the Earth had slowed down the Moon's rotation, just as the Moon's tidal effect on the Earth had slowed down the Earth's rotation.

So it was that after the year of the first lunar landing, 1969, passed, the discoveries of Apollo 11 had accomplished what so many great discoveries do. They had raised more questions than they had answered.

But something else new had emerged—a new science of lunar studies. It was the forerunner of extraterrestrial planetary investigation on a technical and intellectual scale never before envisioned. In a real sense Tranquility Base was the gateway to a new geophysical frontier.

Notes

1. Press, Frank, Apollo 11 Principal Investigators Briefing, Johnson Space Center, July 23, 1969.
2. Ibid.
3. Ibid.
4. Latham, G., personal communication, June 7, 1973.
5. Gast, Paul, Lunar Rock Conference, Department 15, Smithsonian Institution, Washington, D.C., 1969.
6. Gold, Thomas, *Science*, Vol. 165, September 26, 1969.
7. O'Keefe, John, a paper given at the Eastern Analytical Symposium of the American Chemical Society et al, November 20, 1970, Washington, D.C.

8. Lyttleton, R. A., *Mysteries of the Solar System*. Clarendon Press, Oxford, 1968.

9. Ibid.

10. Kopal, Zdenek, "Cosmic Influences on the Early History of the Lunar Surface," a paper presented at the NATO Advanced Study Institute on Lunar Studies, Patras, Greece, September 1971.

11. Smith, J. V., et al., "A Petrologic Model of the Moon," *Journal of Geology*, Vol. 78, 1970, pp. 381–405.

12. Kuiper, G., *Proceedings of the National Academy of Sciences, U.S.*, Vol. 40, 1096 (1954).

13. Ibid.

14. Gast, Paul, "Chemical Composition and Structure of the Moon," a paper given at the Conference on Lunar Geophysics, October 18–21, 1971, Lunar Science Institute, Houston.

5

Ocean of Storms

Eight hours into the second lunar landing mission, the Earth appeared to be the size of a volleyball to the crew of Apollo 12. On the dark side, the three astronauts could hardly tell where the Earth stopped and space began. At 11 hours out, they could not discern any rotation. The home planet just seemed to hang there in black space. The subsolar point (under the Sun) was on water, causing a glint on the surface "somewhat like a light on a billiard ball." After 32 hours, the Earth had shrunk to the size of a golf ball held at arm's length.[1] Apollo 12 was on its way to the Moon.

The crew consisted of three Navy fliers who were members of the most exclusive organizations in aerospace technology, the Society of Experimental Test Pilots and the Corps of Astronauts. The skipper was Navy Commander Charles (Pete) Conrad, Jr., 39, a wiry, balding pilot who had entered the Navy after graduation from Princeton University in 1953. He had received a Bachelor of Science degree in aeronautical engineering. Agile, energetic, and decisive, Conrad was an experienced space man. He had flown on the 8-day Earth orbit mission of Gemini 5 as copilot with Leroy Gordon Cooper, one of the original Mercury 7 astronauts in August 1965. A year later, Conrad commanded the 3-day flight of Gemini 11 with Navy Commander Richard F. Gordon, Jr. On this mission, Conrad maneuvered the Gemini spacecraft into a first orbit rendezvous with an Agena rocket. This was a "first" in space. Using the Agena's rocket thrust, Conrad boosted the joined vehicles to an apogee of 850.5 miles, a new altitude record. Gordon then performed a space-walk on which he tethered the Gemini to the Agena. It was a test to determine whether two vehicles could be stabilized in orbit when linked together only by the length of nylon rope. The experiment was inconclusive, but the flight demonstrated a rendezvous and docking capability that would be critical on lunar voyages.

Now Conrad and 40-year-old Gordon, an efficient and skillful fighter and test pilot, were teamed up again on the second of the seven landfall voyages of Apollo. The third crewman was Navy Commander Alan L. Bean, 37. He was a Navy ROTC student at the University of Texas, was commissioned upon graduation in 1955, and became a test pilot. This was Bean's first trip into space. He was the pilot of the lunar module, which the crew had christened, in Navy tradition, the Intrepid. As LM pilot, Bean was to walk on the Moon with Conrad, while Gordon remained "upstairs" in lunar orbit as pilot of the Apollo command module, Yankee Clipper.

Apollo 12 was launched from the Kennedy Space Center, Florida, at 11:22 A.M., Eastern Standard Time, November 14, 1969. It was bound for Oceanus Procellarum, the Ocean of Storms, on the western side of the Moon. The landfall was a cratered area at 2.94 degrees south latitude and 23.45 degrees west longitude, about 830 nautical miles west of Tranquility Base. The touchdown site was near a shallow crater on the slopes where Surveyor 3 had landed on April 20, 1967. Conrad and Bean were to bring the Intrepid down to a point close enough to Surveyor 3 so that they could hike over to it in their bulky Moon suits, photograph it, and retrieve its television camera and other parts for later analysis at Houston on the effects of two years of exposure to lunar temperatures and the solar wind.

The accomplishment of this assignment required a pinpoint landing from lunar orbit; one considerably more precise than the first landfall by Armstrong and Aldrin the previous July, when the Eagle had overshot the touchdown point by 4 miles. If Conrad and Bean touched down more than 600 feet from Surveyor 3, mission rules would not allow them to attempt to reach it, for 600 feet was the maximum radius they would be permitted to walk from the lunar module.

Scientific Exploration Begins

Given the unknowns about the Moon's "lumpy" gravitational field and the lack of experience in piloting a rocket vehicle to a landing on an airless planet, the odds against achieving a precise landing seemed high indeed. But the flight planners at the old Manned Spacecraft Center, now the Johnson Space Center, had learned a great deal from Apollo 11 and the reports of Armstrong and Aldrin, who were now watching the progress of Apollo 12 like anxious parents.

Apollo 11 essentially had been a test of the Apollo lunar landing system, which called for shuttling to and from the surface of a planet from an orbital "mother" ship. Was this to be the design for future landings on Mars? Most aerospace engineers and theoreticians in the United States thought so and, indeed, had programmed future space exploration for the remainder of the twentieth century on that basis.

What Apollo 11 had demonstrated was the feasibility of the system and the capability of the Moon suits and portable life-support apparatus to keep the explorers alive and reasonably comfortable in the lunar environment. It was axiomatic that if men could land on the Moon and perform useful work there, they could certainly explore Mars—if (and in 1969 it was a big IF) they could endure months of weightlessness in zero gravity without serious and irreparable cardiovascular and bone deterioration. To some extent, the physiological effects of long flights could be projected from the Apollo and Gemini experiences, but, of course, much longer space tours than 14 days would be necessary to see the long-term effect of zero gravity. Such tours were programmed in the orbital workshop (Skylab) project to follow Project Apollo which was in preparation.

Apollo 11, then, was primarily an engineering test of an interplanetary landing and life-support system. It was only secondarily and somewhat incidentally a voyage of scientific exploration.

Having determined that the landing and life-support systems worked well, the Manned Space Flight Directorate in NASA gave full rein to scientific experiments in Apollo 12. Thus, the expedition of Conrad, Gordon, and Bean was the first on which the major objective was the scientific exploration of the Moon.

The lunar module Intrepid carried a complete Apollo lunar scientific experiment package (ALSEP), designed and built by the Bendix Aerospace Systems Division. ALSEP consisted of an array of highly sensitive seismometers that would record the vibrations caused by Moonquakes and meteoroid impacts; a magnetometer that would measure the strength of a magnetic field, if any existed, on the surface; a solar wind spectrometer capable of measuring the energy and direction of charged particles in the solar wind every 30 seconds*; an ionosphere detector designed to measure the rate at which ions (charged particles) formed from the LM exhaust gases escaped from the lunar surface; an atmosphere gauge that would record the composition of gas molecules emitted by any possible volcanic processes; and a dust detector to mea-

---

* Not to be confused with the Solar Wind Composition Experiment, a strip of aluminum foil first deployed on Apollo 11 and also carried on Apollo 12.

sure the accretion of dust on the instruments. The data recorded by this array were to be radioed to Earth on command from Houston from a central station transmitter. Electric power and heating for all the instruments were to be provided by a nuclear electric power system called SNAP-27 (systems for nuclear auxiliary power) developed by the U.S. Atomic Energy Commission. The small power station was a cylinder 18 inches high and 16 inches in diameter. It was constructed of beryllium and weighed 43 pounds (on Earth) when fueled with 8.36 pounds of plutonium-238. The plutonium was carried in a separate lead-lined cask. Radioactive decay of the plutonium produced 1480 watts of heat that was converted into 63 watts of electricity by an assembly of 442 telluride thermoelectric elements. The generator had no moving parts and was designed to last one year.*

The Quiet Moon

In analyzing the structure of the Moon, the seismic experiment was the most important. It consisted of four instruments, two recording vertical and two horizontal motion of the surface. The 21-day performance of the seismometer left on the Moon at Tranquility Base by Armstrong and Aldrin, although initially confusing, had indicated that, compared to the Earth, the Moon was relatively quiet. The Apollo 12 seismometers were extremely sensitive, designed to detect motions of 0.3 millimicron in amplitude. A millimicron is one-thousandth of a micron, which is one-thousandth of a millimeter.

On the eve of the launch, Gary Latham, geologist investigator from the Lamont-Doherty Geological Observatory, described the tentative conclusions he and his colleagues had reached from the Tranquility Base seismic records.[2] Only a few confirmed Moonquakes were reported by the Apollo 11 seismometer during its 21 days of operation. In this period, 200 to 300 events would have been recorded on Earth by such a sensitive instrument. Since internal movement is produced by heat that is generated by the decay of radioactive elements, the paucity of Moonquakes indicated that the Moon was relatively cool compared to the Earth.

Latham said that his group of experimenters surmised that the Moon does not have a large molten core, but it may have a small one. This conclusion was utterly rejected by scientists who believed that the Moon had been formed cold and had never become very hot during all of its

* It was still operating on November 15, 1972, the AEC reported.

history. Such a body would not have a core at all; it would have been too rigid to permit heavier elements (iron, nickel) to move down toward the center, as apparently they did in the Earth.

Latham and his colleagues admitted, however, that the seismic evidence from Tranquility Base did not suggest the existence of dynamic processes such as mountain building or sea-floor spreading, at least not at the rate they occur on Earth. Otherwise, many more Moonquakes would have been seen. Tranquility Base was literally a tranquil place.

On Apollo 12 the seismometers would operate continuously, instead of only during the lunar day as the Apollo 11 experiments did. Also, they would be separated electrically from the other instruments to eliminate interference.

Concerning Apollo 12, Latham observed: "We are really taking quite a step in achieving better science on this mission." The Apollo 11 experiments were just a rather hastily conceived substitute for the complete ALSEP and really a credit to NASA and the Bendix organization in that they were brought about at all in time, he said, adding: "But Twelve [Apollo 12] now really represents the full system and what we have really been pointing toward."

Hy Brasil

As the second expedition crossed a boundary to enter the predominant sphere of lunar gravitational influence, about 200,000 miles from Earth, and began falling Moonward, the expectation of a resolution of the hot Moon–cold Moon controversy was revived among the scientific investigators at Houston by the imminence of new discoveries. But for all their theorizing and attempts to construct hypotheses about the nature of the Moon from the limited data thus far, the scientists of 1969 seemed to be as far away from an acceptable statement of its origin and evolution as were fifteenth-century European mariners from a conception of the Earth's Western Hemisphere in the previous age of discovery. Five hundred years after the Norse voyages to Newfoundland, Europeans were still probing westward in the Atlantic in hopes of reaching Japan and China, unaware that a hemispheric landmass rose in the way and a greater ocean lay beyond. To John Cabot, who rediscovered Newfoundland (for Europeans) in 1497, the way to Cathay lay just beyond, through the fabled Northwest Passage. The fourth voyage of Columbus in 1502 was a search for such a strait from the Caribbean to the Indian Ocean, the naval historian, Samuel Eliot Morison, tells us.[3]

These geographical misconceptions of the age of exploration that preceded ours were no more naive than some of the geophysical ideas concerning the nature of interplanetary medium in 1957 and the nature of the Moon in 1968–1969. Fifteenth-century savants and mariners believed in the existence of such mythical islands as Hy Brasil (Isle of the Blessed) somewhere off Ireland and fabled Antilia, with its Seven Cities, somewhere in the South Atlantic. Twentieth-century astronomers and physicists considered the possibility that a man landing on the Moon was in danger of drowning in deep dust or that the lunar surface was crusted over with fairy castle structures through which a landing vehicle would plunge to its doom in a deep crevasse. These ideas in our own time are hardly less imaginative, let us say, than the medieval speculation that if one sailed far enough, he would be swept right over the edge of the world into some bottomless pit of the nether regions.

In hindsight, many hypotheses and speculations appear incredibly simple, but all through history they have proved highly stimulating to exploration. So they were in Apollo.

The Lightning Strikes

One particular advantage the crew of Apollo 12 enjoyed over explorers of earlier centuries was that they could see where they were going before they left the spaceport. Oceanus Procellarum covers a large part of the western half of the lunar near side. It has an area of about 2 million square miles, considerably larger than the Mediterranean Sea. Most of the mare regions on the Moon are circular, but the Ocean of Storms is not. It contains the largest expanse of mare material on the Moon that is not obviously controlled by a single, circular, multiringed basin, the U.S. Geological Survey has observed.[4] Oceanus Procellarum sprawls irregularly, like a splotch of solder.

Patrick Moore has suggested it may represent lava overflow from the Imbrium Basin to the north into a depression left by an older formation, now nearly obliterated.[5] The U.S. Geological Survey states that the shape of the "ocean" appears to be controlled by the outer ring structure of the Imbrium Basin to some extent. Within this vast, lava-covered expanse are islands of "terrae," or whitish, highland material, and vestigial craters, like the Flamsteed Ring, indicating that the depth of the lava flow is not great. The most imposing feature of the region is, of course, the Crater Copernicus.

Among Apollo scientific investigators, there was a general feeling of anticipation that somehow the lightning of understanding would strike on this mission. The curtain was going up on the manned investigation of the Solar System. Thomas O. Paine, the NASA administrator, shared this view, for he remarked later: "Our second journey to the Moon opened the new age of extraterrestrial scientific exploration by man."[6]

The lightning did indeed strike, not at the end of the voyage, but at the outset. At 36.5 seconds after it was launched, Apollo 12 was struck by lightning as it passed through 6,000 feet altitude. Alarms were set off in the command module as the fuel cell batteries were automatically disconnected by overload sensors and storage batteries picked up the 75 ampere load.

The stroke burned out four fuel gauge sensors in the reaction control system of the Apollo service module and sent the inertial measuring unit, the brain of the guidance system, into a spin. A second bolt of lightning hit the vehicle at 52 seconds after the launch, as the Saturn 5 rocket was thundering up through 15,000 feet. Except for the fuel gauge sensors, which were knocked out, the effects of both strikes were transient. The crew quickly realigned the inertial measuring unit and brought the fuel cell batteries back on line.

At both the Kennedy Space Center in Forida and the Johnson Space Center in Texas, NASA officials were in a dither. Mission rules forbade launching a manned vehicle in a thunderstorm. But at the time of the launch, there was no thunderstorm in the whole Kennedy Space Center–Cape Canaveral area. True, it was a cloudy morning, and Apollo 12 rising on a pillar of flame from its Saturn 5 booster vanished quickly into the overcast. There were broken clouds at 800 feet and a ceiling at 4,000 to 10,000 feet with cloud decks rising to 21,000 feet.

Where had the lightning come from? An investigation ensued. The explanation by Donald D. Arabian, chief of the Apollo Program Office Test Division, turned out to be as fascinating as any event on the flight.

It appeared that the 363-foot Saturn 5–Apollo vehicle and its ionized exhaust plume extending hundreds of feet behind it acted as an electrical conductor between a strong, electrical potential in the low clouds and the ground. The clouds and the ground formed the two faces of a natural and very large capacitor. As Apollo 12 rose between them to first one cloud deck and then another, it discharged the potential in each to the ground.[7] No one had foreseen such a contingency, but once the phenomenon occurred, the hazard was clearly understood. The fact that the vehicle remained hardly scathed when it might have been destroyed was a piece of luck.

Go for LOI

Eighty-three hours later Lucky Twelve approached the point in space where the crew would fire the spacecraft engine to reduce velocity and drop the Apollo, with the lunar module attached like some hapless insect with folded legs, into an elliptical orbit around the Moon. The time was 9:00 P.M., Central Standard Time, November 17, 1969, in Houston. Glynn Lunney, the flight director on shift in Mission Control at that time, polled the controllers at their consoles on a "go" or "no-go" decision for lunar orbit insertion, or LOI. No one had any reason to abort the flight at that point. Apollo 12 was 848 nautical miles from the center of the Moon "motoring along" at a velocity of 6,550 feet per second. Astronaut Paul Weitz, the capsule communicator, called for a report from the spacecraft. Conrad replied that he had made his sextant and star check. The little vessel was dead on course.

"Roger, thank you," said Weitz. "Apollo 12, Houston. You're go for LOI."

"Roger," responded Conrad.

In order to place the vehicle into lunar orbit, Conrad was to burn the Apollo main engine at full thrust for 5 minutes and 52 seconds in the direction of flight. This would slow down the spacecraft by 2,889 feet per second. At this reduced velocity lunar gravity would pull it into an orbit of 62 by 169.3 nautical miles.

The engine would be fired on the far side of the Moon to establish the orbit. At the point of firing, the spacecraft would be out of radio contact with Mission Control. Apollo 12 passed around the western limb of the Moon and controllers at Houston waited. During the interval, the Manned Spaceflight Communications Network chattered briefly around the world over thousands of miles of subsea cable, land lines, and microwave transmission towers.

"Goldstone," called Houston. "You are active. Honeysuckle is the backup. And the spacecraft will come around in normal voice."

Goldstone was the receiving station with a 210-foot diameter antenna at Goldstone, California, in the high desert. It was one of three prime receiving centers for manned and unmanned interplanetary communications. The other two were at Honeysuckle Creek, near Canberra, Australia, and in a suburb of Madrid, Spain. Because of their geographical positions, one or sometimes two stations were always in the line of sight of the lunar near side. Signals from the spacecraft were then relayed by one or more of these stations over the ground and sea links and communications satellites. This technology made it possible to maintain nearly constant radio and television contact with men on the Moon—

the first time this capability had ever existed in the long history of human exploration. Only in lunar orbit was there a gap of about 32 minutes, when the craft passed behind the Moon.

The lunar orbit insertion maneuver was a breathtaking experience for Houston—and had been since the first orbit of the Moon by Apollo 8 the year before. Controllers would not know how accurate the velocity change had been until the craft came around the eastern limb.

After 32 minutes Houston called Apollo 12. No answer. This was indicative that a lunar orbit had probably been achieved. The spacecraft would have been visible a few minutes earlier if its initial wide swing around the Moon remained unchanged. The seconds passed like heartbeats in Mission Control, where all was silent except for a low electric hum in the room and the occasional cough and shuffle of feet of men waiting.

"Apollo 12, Houston. Hello, Apollo 12," the CapCom called.

Static crackled over the loudspeakers. A voice began to filter through, gaining in volume, until it could be heard saying:

"Hello, Houston. Yankee Clipper with Intrepid in tow has arrived on time. Are you ready for the burn report?"

A Poor Liberty Port

The Control Room people exhaled with an audible sigh. Conrad reported that his instruments showed Apollo 12 to be in a 170 by 61.8 mile orbit. Earth radar refined this to 168.8 and 62.7 miles (nautical). Conrad said: "I guess like everybody else that just arrived (nine others had reached this point in Apollos 8, 10, and 11) we are all three just plastered to the window. Looking. For the Navy troops, it doesn't look like a very good place to pull liberty, though. . . . I think these craters are much bigger than anything we have ever seen on Earth."

Bean continued describing the scene to CapCom Paul Weitz:

"We are passing a beautiful impact crater here on our right side now. It's got many, many long rays. It's a beauty."

At that point in Alan Bean's commentary, Mission Control displays showed that Apollo 12 was cruising just 87.1 nautical miles above the ground.

Weitz asked if the crew could see the Crater Langrenus, a landmark.

SPACECRAFT: No, we can't quite see it yet, but we've been looking over at Humboldt [Crater] and at the great fracture marks in it. Actually, it looks to me like some criss-cross roads down there in the desert.

The Apollo television camera began to transmit pictures of the lunar surface. Conrad passed the camera to Gordon so that he could aim the lens at Langrenus.

HOUSTON: Roger. We're picking it up. On the screen, the colors appear to be green to brown. Can you describe the colors in the scene that you're seeing there now?

The color changed as the spacecraft passed over, Bean replied. At first the soil appeared to be a gray-white, like concrete, and then it seemed to acquire a touch of brown. There were several places which looked very white.

SPACECRAFT (Conrad): Another interesting thing is this white or gray-white Moon in contrast, very startling, with the black sky, just like everyone has reported. The black is about as black as you've ever seen in your life. The Moon just sort of very light, concrete color. In fact, if I wanted to look at something that I thought was the same color as the Moon, I'd go out and look at my driveway.

CONRAD (pausing, then continuing): We're passing over the Sea of Fertility now and it's a little darker than the terrain that we've been over, but not much—more a slightly darker gray. Looks like the beach sand down at Galveston, whenever it's wet.

CAPCOM: Roger, Twelve. We had a team of geologists checking your driveway, Pete. We'll send them to Galveston now.

SPACECRAFT: Okay. [Laughter.]

As Apollo 12 began its second orbit of the Moon, the moment of truth approached, the time when Conrad and Bean would crawl through a connecting tunnel from the Apollo command module, Yankee Clipper, into the lunar landing module, Intrepid, and then cast off from the parent vessel to descend to the lunar surface. The procedure was analogous to that of a shore party from a sixteenth-century galleon launching a longboat toward the beach of an uncharted island as the big vessel rode at anchor some distance off shore in deep water. While Intrepid descended to Oceanus Procellarum with Conrad and Bean, Yankee Clipper would continue its cruise in lunar orbit with Gordon at the helm awaiting their return. The landfall was charted only as Site 7. That was where Surveyor 3 was standing, a lone sentinel on the new frontier.

Conrad, Bean, and Gordon continued their description of the lunar surface as it unrolled below them. Each moment, a new vista of the desert world appeared. They were fascinated, but they were pilots, not poets. They did not verbalize the awe and wonder this strange, new, empty land evoked in them. They moved the television camera from one window to another to record the extraterrestrial scene, but the little camera scarcely did justice to it. The crewmen had a word for the view. It was "fantastic," they kept saying.

Morning on the Moon

Oceanus Procellarum looked like a desert, but the astronauts could not think of any desert on Earth that resembled it. The surface was generally smooth, except for occasional long ridges and isolated hills. At first sight the hills looked like cumulus clouds, illuminated on top as they were by the low sun of the lunar morning. The sun was hardly more than 5 degrees above the horizon.

In the beginning of planetary exploration, Earthmen always landed on the Moon in the morning of the lunar day, which is 14 Earth days long. Thus, they were always launched at the New Moon. The reasons for this were practical: The low Sun angle enhanced light and shadow enabling crewmen to clearly see the holes, pits, and craters in the ground over which they were flying and the area where they were trying to land. Low Sun also lent higher photographic contrast. In morning sunlight, the temperature was milder, a mere 50° to 80°F compared to 270° at high noon. It was more feasible to work out of doors in the morning, when the demands on the Moon suit's cooling system were minimal. Finally, the shadow cast by a man or the lunar module provided a means of gauging distance in a landscape where perspective was all askew, where the horizon was all wrong, and where the harsh brilliant glare of daylight shared a knife-edge boundary with the deep and utter blackness of shadow. No sensible atmosphere intervened to diffuse the light. No tree, no bush, nor blade of grass entered the perceptual field of the astronaut to give him perspective. Beyond the undulating Moonscape, with its intermittent ridges and near horizon, only one familiar object was visible to provide a visual reference. It was the blue-white ball of Earth hanging like a mottled pumpkin in the black sky. It told him only that he was one hell of a long way from Houston.

As Apollo 12 passed the terminator where the day ends and night begins, the crew found the lunar surface illuminated richly by Earthshine. In Earth's light, the Moonscape appeared to be gray-green in color, with its contours soft and flowing, and very beautiful. And then around the Moon again they went, to far side, which Earth inhabitants can never see. This region was first photographed in 1959 by the Russian reconnaissance satellite Lunik 2; it is strangely different from near side, like the opposite side of a coin. For the most part, the dark, smooth maria were missing on far side, where great impact scars seemed never to have healed.

In daylight the lunar highlands, or terrae, that sixteenth-century savants thought of as continents were whitish, as they appear from Earth. Also they seemed to be covered with rounded knolls. The highlands were considerably rougher than the maria and more densely cratered. The crewmen reported several areas of vents and cracks in these regions that indicated possible volcanism. Most of the craters, however, appeared to be the result of impacts. One had spewed forth a blanket of ejecta to a distance of 50 to 60 times the crater's diameter. Two craters in the mare displayed perfectly circular ray patterns.

If one dropped a pebble in dust, the cratering effect and the splash-out of dust particles would resemble these lunar craters. There seemed to be no question at all about the bombardment process that had formed this shell-shocked landscape. One might wonder about the origin of the projectiles. What process did they represent in the evolution of the inner Solar System?

Thy Rocks and Rilles

In the great craters there were mountains that looked like volcanoes. These central peaks certainly appeared as though they had ejected molten lava. They were typical volcanic structures. On the central peak of Langrenus, the crew saw boulders. Where had they come from? They saw boulders also on the central peak of the big Crater Theophilus. On that peak ridge lines were visible and quite well defined. Terraces ran parallel to them, whereas rilles, which looked like cracks in the surface, ran perpendicular.

These scenes were evidence of powerful forces at work. They evoked scenes of massive projectiles as big as Mediterranean islands crashing into the Moon, splashing gouts of soil and rock for miles above the lunar plain to rain down tens or hundreds of miles away from the site of impact. Lava flowed from these massive wounds and puddled in the lower reaches. Was it the melt of rock liquified by the heat of impact? Or had the missile punched a hole in a thin crust overlying a region of molten rock?

Two long, parallel rilles excited the crew's attention as Apollo 12 passed over the Sea of Serenity. These large rilles were actually long, narrow valleys. Parallel to them were segments of the ground lower than surrounding territory that are called graben on Earth. They are bounded by faults or big cracks marking the edges of the depressed area. The crewmen saw the fault boundaries quite clearly.

The rilles, the fault lines, and the graben were indicative of an active planet, not a dead one, as many scientists supposed the Moon to be. They were unmistakable evidence of tectonic (mass-moving) activity. It spelled heat energy within the Moon. Perhaps these features had been formed far in the past, but they were clues from which a molten Moon, or at least a partially molten Moon, could be deduced.

As Apollo 12 passed around the far side of the Moon to begin its third revolution, the crew prepared for another engine burn. It would insert the vessel into a lower orbit of 66.3 by 54.7 nautical miles. From this orbit, the Intrepid would commence its descent to the surface. Its target, Site 7, was marked by three small craters so aligned as to resemble a crude snowman. In the center, or "belly" crater of the snowman, stood Surveyor 3 and, naturally, this crater was called Surveyor Crater, whereas the one just west of it—the head of the snowman—became Head Crater for purposes of identification.

At 2:10 A.M. on November 18, 1969, Conrad fired the Apollo main engine for 17 seconds in the direction of flight. The thrust braked velocity by 165.5 feet per second. The vessel sank into a new orbit of 66.3 by 54.7 nautical miles.

Through the early morning hours of November 18, the crew prepared the Intrepid for the descent to the Moon. At 7 A.M. the astronauts began their 8½-hour rest cycle. While all was silent aboard Apollo 12, Mission Control was notified by the Solar Particles Alert Network that a moderate flare had been observed on the Sun by stations in the Canary Islands and Africa. It was not a big flare, but it heralded the onset of a solar storm of high-energy particles (protons and electrons) that would race past the Earth and Moon in 4 days.

Normally, these subatomic bits meandered through the Solar System at gentle velocities with the solar wind, but, occasionally when they were accelerated to high velocities, the effect was more like that of a hurricane, a solar hurricane. On Earth, these particles streamed into the atmosphere at the magnetic poles, interacted with the ionosphere, the electrically charged portion of the upper atmosphere, caused communications blackouts, and brilliant auroras. For the most part, however, the surface of the Earth was shielded against the high-energy particles by the umbrella of its great, magnetic field, extending 40,000 miles out in space on the Sunward side of the planet.

The Moon, however, orbited outside the Earth's magnetic umbrella. With no appreciable magnetic field of its own, nor atmosphere, its surface lay exposed to the full impact of the particle storm. Astronauts on the surface during such a storm might be subject to exposure to higher-level radiation than their suits could ward off. No one knew how high the levels might be on the Moon. Thus, the solar flare monitoring network served as an early warning system. Far above the Earth orbited a number of military and NASA satellites that would report the intensity of the ensuing particle storm. If the storm was severe, the astronauts would remain indoors until it abated. But, as the flare progressed, it became evident that a severe storm was not in prospect. So far as solar weather was concerned, the prospects were good that it would be fair.

It was hard to believe in a windstorm of nuclear particles, of protons, nucleons, or electrons, but Moon-bound astronauts had experienced the visual sensation of bright flashes before their eyes when their eyes were closed. The flashes, they learned, were caused by the impact of high-energy cosmic-ray particles striking the optic nerve! The particles tore through the walls of the ship. They zipped through bulkheads as through a net. They were both the mote and the beam in the eye. They pierced the brain. Once a man had experienced the flashes, he no longer regarded particle radiation as an abstraction. The flashes made him a believer.

The Second Descent

Meanwhile, Alan Bean had complained of a "stuffy head" during the flight. If a head cold was coming on, it might cause him serious difficulty on the surface. Mission Control recommended a decongestant. Conrad was sure that Bean's trouble was caused by dust in the cabin. There was a lot of dust and it was bothering the crew.

After the sleep period ended, Conrad called down to Mission Control: "I'd like to square something away down there. Al doesn't have a cold and all I have is a one-inch itch and I don't consider it any big, major medical problem. As a matter of fact, we're in pretty damn good shape."

CapCom acknowledged: "Roger, Pete. Just talked to Jane (Mrs. Conrad) a few minutes ago. She said the family's all doing well and they're all getting rested up to spend the night watching you."

The Conrads were the parents of four boys ranging in age from 8 to 14. Gordon and his wife Barbara had six children, two girls, 8 and 15, and four boys, 9 to 14. Bean and his wife Sue had a boy, 13, and a girl, 6. The families were all set to watch the second descent of man to the Moon on television.

Preparations continued during the later afternoon of November 18. Mission Control wanted photos of the Fra Mauro formation, the target for the next mission, Apollo 13.

At 7:09 P.M., CST, Conrad and Bean extended the lunar module's landing legs, which had been folded during the flight like an aircraft landing gear. They undocked the Intrepid from Yankee Clipper at 11:16 P.M. The Clipper briefly burned its small reaction control thrusters to move away from the Intrepid, so that the lunar lander would have space room to maneuver.

"Bye, bye," called Conrad over the radio as the mass of the Apollo command and service modules drifted away from the tiny Intrepid.

"See you, troops," replied Gordon.

At midnight the glass enclosed viewing room behind the consoles at Mission Control was filled to standing room only with VIPs, including Administrator Paine; his deputy, George M. Lowe; astronauts Armstrong and Aldrin; astronaut Frank Borman, who commanded the first Apollo to orbit the Moon (Apollo 8); C. Stark Draper, director of the Instrumentation Laboratory of the Massachusetts Institute of Technology, where Apollo's inertial guidance system that was revolutionizing aircraft navigation had been developed; and Wernher von Braun, then director of the Marshall Space Flight Center. Not a single member of the news media, however, was present. Newsmen were barred from Mission Control as a bureaucratic safeguard against uncontrolled disclosure in the event of a disaster. This policy had been established in Project Mercury and it was continued until late in Apollo, when a press pool representative was finally admitted to the viewing chamber behind the control room at the Johnson Space Center.

Program 63

A landing on an airless planet required continued thrust against the planet's gravitational pull. The LM descent engine, with a maximum rated thrust of 9,870 pounds, had proved its capability of letting the vehicle down gently on Apollo 11, reducing velocity from 5,560 feet per second to zero at touchdown.

As the detached vehicles vanished around the far side of the Moon on their fourteenth revolution, Gordon maneuvered Yankee Clipper into a nearly circular orbit of 60 nautical miles. Conrad and Bean, standing up in the Intrepid like charioteers, burned their powerful descent engine to reduce velocity by 72 feet per second. This descent orbit initiation burn put the LM into a flight path that dipped to 50,000 feet from the surface at a point on near side about 15 degrees uprange of Site 7.

At this low point, or perilune, of the new LM orbit, the crew had the option of commencing powered descent to the surface or, if any problem intervened, of simply rounding the Moon again—or even rejoining Yankee Clipper in the event of a breakdown.

A few minutes after midnight, as November 19 came to Houston, Intrepid reappeared, with Yankee Clipper some distance away. Conrad reported: "A great DOI (descent orbit initiation)."

The Intrepid was flying on a long, curved trajectory that would bring it to within 50,000 feet of the ground. All was going well.

Houston announced: Thirty minutes to PDI (powered descent initiation).

ALAN BEAN (gazing out the window at the lunar horizon and seeing the Earth): You're about 30 degrees above our horizon, now, Houston. And you're about a one-third crescent Moon. And you really are beautiful. Big, blue and white.

CAPCOM: Roger. We put on our Sunday best for you.

BEAN: We're pretty well suited out up here ourselves. Boy, this thing sure flies nice.

At 12:33 A.M. on November 19, the Intrepid's guidance system executed Program 63, an automatic maneuver, positioned the little ship for powered descent initiation facing the sky, with the windows up and the engine bell forward.

As the Intrepid coasted downward toward 50,000 feet and PDI, Flight Director Clifford Charlesworth polled controllers for a "Go" for the initiation of powered descent. It was unanimous. It was "Go."

CAPCOM: Intrepid, Houston. Go for PDI.

INTREPID (BEAN): Prepare descent in 35 seconds.

INTREPID (CONRAD): It'd do it in 30, Al.

INTREPID (BEAN): Okay. The average G (gravity) descent engine is on and the velocity light is on. Got it made! A couple of pretty lights here! I have ignition.

CAPCOM: Throttle up to 26 (maximum thrust).

INTREPID: Yep. Feels good to be standing in a G field. [This was provided by the engine thrust against free fall.]

The lunar module Intrepid was coming down on the Moon. But the crewmen could see little but sky and a little horizon until they reached a point called "high gate" at 7,335 feet. Then the craft was rotated so that the windows faced forward and the crewmen could see the ground.

As Conrad stood by the controls, Bean called out velocity changes from the guidance computer. The Intrepid dropped rapidly through 45,000 feet, then through 35,000 feet. At 19,000 feet, Conrad said: "I've got some kind of horizon out there. I've got craters, too, but I don't know where I am." The vehicle was arcing forward so that the windows began to reveal a distant surface.

CAPCOM: Intrepid, Houston. You're looking good at 8. [Eight miles up-range from the landing site.]

INTREPID (CONRAD): That's 12,000 feet according to our tape meter, Houston. I'm trying to cheat and look out there. I think I see my crater. . . . Hey! There it is! Son-of-a-gun, right down the middle of the road. . . . We are passing 3,500 . . . coming down at 99 feet per second. Looking good. Got 15 percent fuel.

CAPCOM: Intrepid, Houston. Go for landing.

INTREPID: One thousand feet. Coming down at 30. Got 14 percent fuel. Looks good out there, babe. Looks good.

INTREPID (BEAN): Thirty-five degrees, 530 feet, Pete. 530. 471. All right. 426.

INTREPID (CONRAD): All right. I got it.

INTREPID (BEAN): Four hundred. You're at 366, Pete.

INTREPID (CONRAD): Right.

INTREPID (BEAN): Oh, look at that crater! Right where it is supposed to be. You're beautiful—257 feet. Coming down at 5. [Five feet per second]. Ten percent fuel. Two hundred feet, coming down at three. You can come on down.

Contact!

INTREPID (BEAN): Forty-two feet. Coming in 3 [feet per second], coming down in 3. Okay, start the clock. Forty-two feet, coming down in 2. Forty, coming down in 2. Looking good. Watch the dust! Thirty-two, 31, 30 feet. Coming down in 2. Pete, you got plenty of gas, plenty of gas, babe. Stay in there.

CAPCOM: Thirty seconds.

INTREPID (BEAN): Eighteen feet, coming down in 2. He's got it made. Come on in there. Contact light!

A blue light flashed on the Intrepid's console showing that extended "feelers" from the landing pods had made contact with the lunar soil.

CAPCOM: Roger. Copy. Contact.

INTREPID (CONRAD): Okay, ignition off. I cycled these valves. Ascent (engine) expendables look good. Okay, we are in hot shape, Houston. We are in real good shape.

One of Conrad's first checks on landing was to make sure that the fuel supply was intact for the ascent engine which would lift them off the Moon to rejoin the Yankee Clipper after their 31½-hour stay on the surface.

It was 12:54 A.M., CST, on November 19, 1969. The second manned lunar expedition had landed in the Ocean of Storms 110 hours and 32 minutes after liftoff from Florida.

The Ocean of Storms presented a different aspect from the Sea of Tranquility where the first crew had landed.

CONRAD: Man, oh man, Houston, I'll tell you I think we're in a place a lot dustier than Neil's. Good thing we had a simulator [on which to rehearse the landing] because it was an IFR [instrument] landing.

In the lunar module Eagle, Armstrong and Aldrin had been able to see the ground as they approached and skipped the vehicle over a field of boulders. But the cloud of dust raised by the Intrepid's descent engine interfered seriously with visibility by Conrad and Bean.

Soon, however, the dust settled. The explorers peered out the window.

CONRAD: Holy Crane. It is beautiful out there.
BEAN: It sure is.

Passing overhead in the Yankee Clipper, Gordon called down his congratulations for the safe landing.

INTREPID: Thank you, sir. We'll see you in 32 hours.
CLIPPER: Okay, have a ball.
INTREPID (BEAN): It's a nice place to land. Look at those boulders out there on your right, Pete. Gee, they're big. It's a pretty good place.

From their perch in the ascent stage of the lunar module, the third and fourth humans to land on another planetary body, as the NASA commentary referred to Conrad and Bean, looked out upon an undulating plain. It was all rolling country with no mountains, hills, or scarps. There were blocky boulders strewn about the ground and little color contrast. The ground looked white in the sunrise.

INTREPID (CONRAD): Those rocks—are we on the Copernicus ray area?

He referred to the splash-out from the big Crater Copernicus, 600 miles away. From Earth and also from lunar orbit, the splashed out rocks and soil appeared lighter in color than the surrounding ground and formed a raylike pattern.

Conrad then confessed to Houston that he shut off the descent engine a bit too soon.

CAPCOM: Shame on you.

CONRAD: Well, I was on the gauges. That's the only way I could see where I was going. I saw that blue contact light and I shut that baby down and we just hit from about 6 feet.

CAPCOM: Roger, Pete. The Air Force guys say that's a typical Navy landing.

With Conrad and Bean in Oceanus Procellarum

Between July and November 1969, confidence in what men could do on the Moon had escalated remarkably. The occupants of Eagle, Armstrong and Aldrin, had spent only a short time out of doors. The total span of their extra-vehicular activity was 2 hours, 32 minutes. It was planned that Conrad and Bean would spend at least 7 hours outside, with an extra half-hour thrown in if all went well. This period would be divided into EVAs of 3½ hours each, separated by a 9-hour sleep period.

The Apollo 12 explorers would spend 31½ hours on the Moon altogether, compared to 21 hours, 36 minutes spent by their predecessors. Total stay time in lunar orbit would be increased from 60 hours (Apollo 11) to 89 hours, or 45 revolutions of the Moon.

The radius of exploration on foot would be extended from Neil Armstrong's stroll of 250 feet in July to about a half-mile, assuming a walking rate of 2½ miles an hour, which would give the explorers time to get back into the LM on emergency oxygen in case of failure in the portable life-support system they carried on their backs.

During the initial 3½-hour EVA, Conrad, first down the ladder, would collect a contingency sample of Moonsoil and then deploy the modular equipment stowage assembly (MESA). This was a tablelike structure that unfolded from the base of the lunar module and held rock sample containers, tools, and the television camera. Inside the LM, Bean would turn on the TV camera and in addition take movie camera pictures of Conrad's activities.

Meanwhile, down below, Conrad would take an equipment bag out of the MESA, plus a set of spare portable life-support batteries and a cannister of lithium hydroxide (an air purifier). Loading these in the bag along with the contingency sample, he would pass the bulky package up to Bean in the cabin by means of a pulley cable that was installed on the module to enable the explorers to transfer equipment about 12 feet from the cabin to the ground.

When these articles had been passed up to the cabin, Bean would lower two 70-millimeter Hasselblad cameras in the bag and then descend to the ground. While Bean was getting his Moon legs, Conrad was supposed to be opening the umbrellalike S-band antenna that was to relay television pictures of their activities to Houston.

Bean then would set up the aluminum-foil solar wind collector that would be exposed a total of 30 hours, compared to barely more than 2 hours on Apollo 11. Both explorers were then to move counterclockwise around the LM, taking both panoramic and close-up photos—the close-ups to show how the LM landing pods had impacted the soil.

All these activities were carefully plotted out in advance, and the explorers were expected to follow the plot as faithfully as actors in a play in order to maintain the "time line" of progress. Only the dialogue was their own in this real-life drama. No such script of stage directions had ever been contemplated in terrestrial exploration, but the Moon was a new environment into which man was extending a tentative toe. Every movement was to be planned, timed, recorded, and every unplanned move justified, explained. It seemed at times as though the explorers were little more than puppets manipulated from an electronic string a quarter of a million miles long. For the entire mission, virtually every event, every movement, was governed by a script as thick as a telephone book—the flight plan.

Having set up the television camera at the end of its 90-foot cable so that it would show everything they did, Conrad and Bean were programmed to set up the five lunar surface experiments. Bean was to remove the nuclear fuel from its heavy cask and insert it into the thermoelectric generator, which would power the central broadcasting station for the experiments. The whole array was positioned about 100 yards from the LM so that the exhaust from its ascent engine on departure from the Moon would not damage the instruments. The four seismometers, the magnetometer, the solar wind spectrometer, the ionosphere detector, and the atmosphere detector were to be erected in a semicircle around the central station.

When all this was done, Conrad was to signal Houston to command the central station radio transmitter to turn on. After verifying that the station was operating, the explorers were to trudge back to the LM, collecting rocks and soil as they went and driving 15-inch tubes one inch in diameter into the ground to collect cylindrical core samples of the soil.

They then were to climb back into the cabin and hoist up their rock and soil collection—about 34 pounds. When this was completed, the flight plan called for them to close the hatch, sling their hammocks, and sleep for 9 hours. The hammocks were an innovation on Apollo 12. Armstrong and Aldrin had slept on the Eagle's floor.

The second EVA called for them to reach Surveyor 3 if they could, retrieve the television camera, a portion of the camera's cable, some aluminum tubing, and some glass and pack all these items in the equipment bag for return to the Earth. Biologists were particularly interested in the cable on which they had taken bacteria counts before the Surveyor was launched. Could the bacteria have survived the lunar environment for 2½ years? What had happened to the aluminum tubing, the glass, and the TV camera and its parts on the Moon? The results might be fascinating if the astronauts could reach Surveyor.

In the preflight briefings by NASA, no guarantees were offered that Conrad and Bean would even land near Surveyor 3.

Now, in the wee hours of November 19, 1969, Conrad and Bean were on the Moon, looking out upon a whitish desert, facing a rigid and demanding schedule planned long in advance. The only option they really had was failing to carry out their assignments. It was as though these men were the extension of a vast organism whose brain was housed on the Gulf plain of Texas and whose neural impulses moved at light speed on radio frequencies across the cislunar void.

Guess What I See!

"For planning purposes," Conrad reported, "we landed very close to the head of the snowman."

The Public Affairs Officer at Houston announced that the Intrepid had come down about 700 to 800 feet from Surveyor—probably on the rim of Head Crater, just west of Surveyor Crater.

At 4:38 A.M., CST, the Yankee Clipper radioed good news. Gordon, looking down through his 28 power sextant glass as the Clipper passed over, spotted the Intrepid on the slope of Surveyor Crater itself, hardly 600 feet from Surveyor 3.

Conrad and Bean, with a great deal of assistance from Houston, had scored the greatest navigational feat in history. They had made a pinpoint landing on the moon.

HOUSTON (Shortly after 5 A.M.): Intrepid, you are go for EVA.

Conrad and Bean helped each other into their Moon suits. Conrad backed out the hatch and got his feet on the ladder:

BEAN: Hey, just a second. Don't go down yet. I've got to get my camera on you, dude.

CONRAD: I can't go down yet, anyhow.

He had become tangled in the conveyor line that had been slipped out of the hatch—the rope and pulley system on which the explorers

APOLLO 12 TRAVERSES

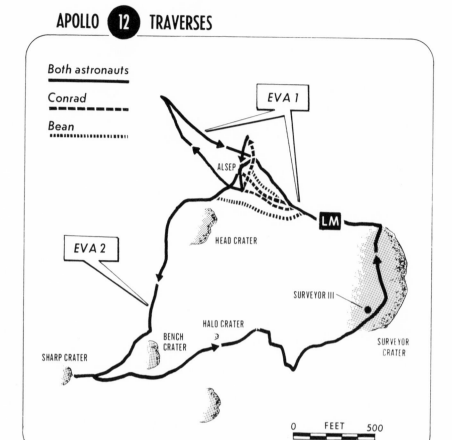

would transfer cameras, tools, and rocks between the cabin and the ground.

Conrad untangled himself and proceeded down the ladder, breathing a bit rapidly. At last, he stood on the LM landing pad, looking up at the ungainly vehicle that had brought him there.

CONRAD: Man, is that a pretty looking sight, that LM. Whoopie! Man! That may have been a small one [step] for Neil but that's a long one for me.

[That was the last step on the ladder Conrad was talking about. Even the 5-foot 11-inch Armstrong remarked that it was a big step for him. It was a bigger one for the 5-foot 6½-inch Conrad.]

CONRAD: I'm going to step off the pad. Right. Oh, is that soft! Hey, that's neat! I don't sink in too far. Boy, that sun's bright. That's like somebody shining a spotlight on your hands. I can walk pretty well, Al, but I've got to take it easy and watch what I'm doing. Boy, you'll never believe it! Guess what I see sitting on the side of this crater. The old Surveyor! Does that look neat. It can't be any farther than 600 feet from here.

CAPCOM: Well planned, Pete.

CONRAD: Yes—just a couple of months for a lot of people.

Following the flight plan precisely, Conrad scooped up the contingency sample. Bean noted that his partner seemed to be leaning forward too far. He warned Conrad not to fall over. It would be difficult to get up in the Moon suit without help.

CONRAD: Am I really leaning over, Al?

BEAN: You sure are.

CONRAD: It seems a little weird. I don't think you're gonna steam around here as fast as you thought you were.

Conrad then opened the MESA, extracted the equipment bag, the batteries, and the cannister, and, stowing these items in the bag along with the contingency sample, he put them on the pulley line to pass them up to Bean.

Hopping around in the Moon suit, Conrad said, made him feel just like Bugs Bunny.

After Conrad had been outside 40 minutes, CapCom gave Bean a "Go" to climb down to the surface. Conrad warned him to watch it in the shadow "because you can't see what you're doing." When Bean arrived on the ground, Conrad raised his gloved hand and pointed to Surveyor, an artifact of tarnished silver in the morning light. "Beautiful," breathed Bean. "Beautiful."

Both men commented on the shiny glass "icing" they saw on the rocks in small craters. Conrad pointed to samples of pure glass.

"Hey," said Bean, "it's real nice moving around up here. You don't seem to get tired. You really hop like a bunny. Where, oh where, is Earth?"

Conrad told him to look straight up where the S-band antenna pointed.

"There it is," said Bean, staring at the blue-white pumpkin where it was morning in the Western Hemisphere.

Once they set up the television camera, the Moon men expected reactions of awe and satisfaction from Houston with the revelation of the lunar scene. But Houston saw nothing but a blank partial brightness in its screens. Bean was asked to wave his glove in front of the lens to see if it was blind. It was. There was no change in the empty brightness.

The picture, said CapCom, was 20 percent bright at the top and black at the bottom.

Bean examined the lens. He reported that it was focused at infinity with the zoom at 30 feet. The *f*-stop, he said, was set at 22. Leaving the television camera, the Moon men proceeded to set up the surface experiments and plant the United States flag. While visual communication had somehow failed, they were still in close voice and telemetry contact with Houston.

The minutes rushed by, but the explorers stayed doggedly on their time line. At one point Conrad, moving down a slope, executed a slide. He found he could still maintain his balance. One didn't learn to walk on the Moon all at once—it took time, but it wasn't really difficult. There were a number of mounds 4 to 5 feet high that puzzled the astronauts. The mounds looked something like inverted craters.

After setting up the experiments, Conrad and Bean walked to a crater northwest of the LM. They picked up three big rocks plus a bag of fines and filled one bag. They stuffed fines and a dozen rocks in another bag. Altogether they collected nearly 50 pounds of rock and dirt samples, which they hoisted up to the cabin on the rope pulley.

Meanwhile, back at Mission Control, a group of happy seismologists watched the passive seismic experiment readout respond to the footfalls of the men on the Moon. Radio signals from the Central Station showed that the magnetometer was working, too. From the thermoelectric generator, 73 watts of electrical power was flowing into the instruments, more than enough to operate them and to keep them warm at night.

The Moon men completed their work in 4 hours and 1 minute—31 minutes longer than planned. They were not tired and their consumables—the breathing oxygen and water in their suits—were not at low levels. Nevertheless, Mission Control urged them to return to the cabin. Their most difficult task was extracting the plutonium fuel capsule from its lead-lined cask. Bean had to hammer on the base of the cask with the mallet provided for driving core tubes into the ground to loosen it.

A Black Snowstorm

After reentering the cabin and repressurizing it with oxygen, Conrad discussed plans for EVA 2 with Houston. He suggested that they climb down to the bottom of the crater and walk uphill to the marooned Surveyor. The slope where Surveyor rested looked too steep to negotiate safely from the rim behind the spacecraft.

"I think we're pretty well game for any traverse that you want us to make," he said.

A tentative traverse route was mapped out. The total distance to be covered was about a mile. Although this journey was miniscule compared to those of polar explorers Peary, Amundsen, and Scott earlier in the century, it was the first traverse by men on another body in space.

Apollo 12, the second lunar landing voyage, demonstrated that man's expansion into the Solar System depended on his technological ability to master a hostile environment. And this, in turn, depended ultimately on his capacity to understand it. On this effort rested man's fate and his ultimate survival, not merely as a species, but as a civilized being. For unless he could learn how to control or adapt technologically to hostile environments, he could not expect to outlive the changes that had extinguished so many species before him in the long history of the Earth.

Although this underlying motive for space exploration was rarely verbalized by NASA people, it formed a basis for the rationale on which manned space flight was being developed. It imparted a sense of mission to administrators, engineers, astronauts, and scientists that far outweighed the petty, short-sighted, and narrow arguments against the national investment in manned flight.

Later that morning of November 19, the EVA was reviewed at a Manned Spacecraft Center press conference by Armstrong and Aldrin.

"I guess we could just say that from our point of view it was an extremely successful day," said Aldrin. "I didn't see anything . . . that would indicate any significant surprises from the kinds of things that we ran into during our flight."

Conrad and Bean had commented several times about the amount of dust they were encountering. Aldrin said he gathered from their conversation that the dust problem was about the same as he and Armstrong experienced in the Sea of Tranquility.

"Pete complained about dust raining down on them during the time they put the rock boxes up, coming off the pulley. The same thing happened to us. We discussed that with Pete before the flight and I am sure he was aware that it would probably occur, but being in a black snowstorm is always a little bit of an unusual occurrence."

Armstrong said he got the impression that the material was looser than "what we had seen both by virtue of the fact they were able to push the flag in further and get their core tube in much easier." The fresh crater was interesting, he continued. "Just about everything we saw was rounded—no evidence of any recent impact."

Armstrong called attention to the pinpoint landing that his colleagues, Conrad and Bean, had executed. It was significant in terms of future flights, which would require precise navigation. Conrad and Bean had proved, he commented, that men could land anywhere on the Moon they wanted to.

Rock and Roll

After a nap of 5 hours, Conrad and Bean awakened shortly after 7:15 P.M., CST. CapCom had some messages for them, including a request to stow the malfunctioning television camera in a lithium hydroxide cannister with the lens up. The failure of the camera had been a setback. It was suspected at Mission Control that the electronics had been burned out by the inadvertent, direct exposure of the lens to the rising Sun. The technicians wanted to look at the camera when the crew returned in order to prevent a similar television debacle on the next flight.

Meanwhile, as the Intrepid "lay at anchor" in the Ocean of Storms, another vessel of the same name sailed the placid Caribbean Sea, a quarter of a million miles away. It was the U.S.S. *Intrepid*, an aircraft carrier, and her officers and men sent a message of good luck to their ship's namesake on the Moon. For the time being, at least, the United States had a three-ocean Navy.

Mission Control, in the meantime, had revised the traverse. The astronauts were instructed to walk first to Head Crater, photograph it, and document its slope, slump, and ledges. Then they were to take a stereo pair of photos of a rock sitting on the crater rim. Next, they were to roll the rock down the crater slope and photograph it as it rolled. The photos might illuminate this type of mass transfer on the Moon and a seismic "signature" of this event would be relayed to Earth by the seismometer array.

CAPCOM: That's point one. Do you copy?
CONRAD: Yessir. We shall rock and roll.

Next, continued CapCom, the explorers would hike to Bench Crater and then proceed to Sharp Crater (see map on p. 111). They would take stereo pairs of photos of any features of interest in Bench Crater, especially of the bench, a rock table in the crater; determine whether the bench is bedrock or breccia; and, last, from the rim of Bench Crater, look northwest and southwest to see if material believed to be a ray from Crater Copernicus differs from adjacent material. Following a visit to Sharp Crater, they would go to Halo Crater, so called because it appeared to be surrounded by a halo of bright material, and then try to examine Block Crater.

Taking note of this itinerary, Bean reported what appeared to be an astonishing phenomenon. The solar wind composition trap, consisting of a 0.5-mil-thick banner of aluminum foil 4 feet long and 1 foot wide, had been hanging straight down when the astronauts climbed back into the LM earlier that day. After awakening from their sleep period, they noticed that the foil was pushed back against the pole, as though by a brisk breeze. But there is no breeze on the Moon—just the so-called solar wind of subnuclear particles, blowing out through space from the Sun.

BEAN: It looks like a sail in the wind around the pole, bulging in the front and bent back on the sides. Real crazy.

CAPCOM: We've got a real solar wind, I suspect, Al.

BEAN: You may think you're kidding, but . . .

CAPCOM: No, Al. It could be that maybe the front part is thermally expanding a lot more than the back. The back probably is just radiating and the front probably is very hot. A thermal difference across could do it. I'm meeting with a lot of approval back here on that idea.

BEAN: Naturally. Yes, well, it looks like it's wrapped around the pole and that's a funny thing. It looks like the wind is blowing on it.

At 9:49 P.M., CST, November 19, the explorers began their second EVA by depressurizing the Intrepid and opening the hatch. Conrad once more was the first man down the ladder and when he reached the ground, Bean passed the equipment transfer bag with the two 70-millimeter cameras down to him.

The Sun was higher in the eastern sky and Surveyor was more clearly visible now. CapCom asked Bean to advise Conrad that the length of his shadow was 18 feet. This would give him a distance reference. Ask Conrad to photograph the solar wind composition sheet wrapped around the pole, CapCom added.

But as Conrad prepared to take the picture, he saw that the foil no longer appeared to be wrapped around the pole. It must have been an optical illusion from inside the LM, he told Bean, who reported this observation back to Houston.

Bean then came down the ladder and the two moved to Head Crater, where Conrad picked up a grapefruit-sized rock. They took the stereo photos Houston had requested, and then Conrad rolled the rock downhill and they photographed that. Bean was intrigued by a small crater about 3 feet in diameter where he saw glass-tipped rock fragments.

Kicking away the topsoil from the crater rim, Conrad found the under-soil to be lighter in color, like cement.

With a small hand shovel he dug a shallow trench at the northwest corner of Head Crater. The undersoil continued to appear a light gray, compared with the darker gray of the topsoil. "Every little crater you come to," said Bean, "you see the glass beads."

At this point in the traverse, Conrad and Bean had moved in a wide semicircle to the south of the Intrepid.

"Man, does that LM look small back there," said Bean. As they walked a bit farther to the east, Conrad reported that he could see the LM at 10 o'clock, indicating that they were then southeast of the vehicle.

A Giraffe in Slow Motion

Ahead of them stood the patient robot, Surveyor.

CONRAD: You know what I feel like, Al?

BEAN: What.

CONRAD: Did you ever see those pictures with giraffes running in slow motion?

CAPCOM: Say, would you giraffes give us some comment on your penetration as you move across there?

BEAN: Oh, it's much firmer here. The toe sinks in a bit as you push off. Every time you land, it sends little particles spraying out ahead, 2 to 3 feet.

CONRAD: Hey. We're back at Bench Crater.

Looking at a photo map, the Capsule Communicator at Houston gave them directions for reaching Halo Crater after they finished chores at Bench and Sharp Craters. They should go due east, into the Sun, he directed. Halo Crater would look like Bench Crater, but would be only half as wide.

CONRAD: I'll tell you one thing. I'd go for a good drink of ice water.

It was getting hot in the lunar desert as the Sun rose slowly higher into the pitch black sky.

After photographing Bench Crater, they collected samples of a fine and medium-grained igneous rock and some fine-grained loose dirt that they stowed in a sample bag. Then, moving westward, they reached Sharp Crater and photographed it. Bean dug an 8-inch trench and Conrad took a core sample.

Bean reported that his hands became hot when he carried the shovel or other tool. Conrad said his legs weren't tired, but his hands were, from grasping tools in the heavy Moon gloves.

"Tell Jim Lovell to practice digging," advised Bean. Lovell was commander of the next expedition, Apollo 13.

The two then moved off to the east toward Halo Crater, and when they reached an area they though might be this objective, they took panoramic photos and collected a double core tube sample by fitting two of the 15-inch tubes end to end and driving them into the soil. While hammering, Bean noted that little chips of metal were flaking off the hammer. The chips seemed to be a coating of some kind.

"I can't believe we're at the right place," said Bean.

"I'm not sure we're at the right place either," said Conrad. Nevertheless, they worked at getting their samples.

As they walked away from what they hoped was Halo Crater, Conrad picked up a rock with a big, glassy splotch. It looked as though someone had poured molten glass on it. At this point, they had been out 2 hours and 15 minutes and Houston said they could extend the programmed 3½-hour EVA to 4 hours. So they had another hour and 45 minutes to go. It seemed an incredibly short time to spend at a place they had journeyed so far to explore.

"You know," said Bean, "we are collecting a lot of the same kinds of rocks, but there just doesn't seem to be any other kinds around. . . . Haven't seen any microbreccia the whole day. This is not at all like Neil's (Armstrong) run."

Conrad addressed Houston. "We just made a sample of glass beads and some local rock on the south edge of Surveyor Crater, Houston. They are going into bag 14D."

A Tan Surveyor

At Mission Control there was concern that as the explorers entered Surveyor Crater, they would sink into deep dust. The prospect of explorers drowning in dust had been a bugaboo for years, as I have mentioned earlier, and fears that a whole spacecraft might disappear in a morass of dust had haunted the designers of the lunar module. It was not until the first Russian Luna and then Surveyor 1 landed that the Apollo directorate could be at all certain it was possible for a heavier vehicle to land on the Moon. Now there was concern that the explorers might encounter quicksandlike dust at the bottom of the crater.

Houston kept calling for reports on how far the explorers were sinking in, but Conrad reassured Mission Control there appeared no danger of that. The footing remained firm.

Bean reported they were moving radially around the inner slope of the crater toward Surveyor. He advised Conrad that Houston was worried about their getting out.

BEAN: We've been thinking about that, too, Houston.

CONRAD: Don't worry, Houston, because it is really no strain. I'm 200 feet away from it, at the same level. The ground is firm and I can come right back up the way I came in.

BEAN: That's right.

CAPCOM: Roger. Sounds good.

As they neared the Surveyor spacecraft, they decided to "take a little rest." Moving along the inside crater wall was harder work than they expected. It was thirsty work, too.

CONRAD: I could have landed the LM at the bottom of the crater. It would have scared me to death, but. . . .

BEAN: What color is the Surveyor supposed to be?

CAPCOM: It was painted white.

CONRAD: It looks like a light tan now.

BEAN: It changed color, huh?

CONRAD: Sure has. On the slopes here, it's just a little bit softer. There's no tendency to slip down or anything like that.

Surveyor 3 loomed unexpectedly large and solid against the crater rim. For 2 years and 5 months it had been standing on the dusty slope, an alien, Dali-esque structure, a metal Cyclops, with its single television eye staring blindly at the ground. It had become, indeed, a light tan in color. Some parts of it looked brown. Conrad and Bean stood before the robot, peering at it through their visors.

"Boy, it sure dug in the ground, didn't it?" observed Conrad. "Look at those pad marks [from an initial impact bounce]. They're still here. The aft honeycomb shock absorber struck the dirt and it looks like it took some of the shock. Oddly enough, the front one didn't appear to do that."

Bean photographed footpad no. 3 close up and scooped out some soil near it as a sample. Then they examined the Surveyor television system and got the shock of their traverse. The TV field of view was enhanced by a mirror, which, like the rest of the vehicle, had turned a light tan in color. But when Bean rubbed his glove over the mirror, the brown color disappeared and the glass became shiny. The vehicle had looked tan because it was completely coated with a fine layer of dust—space dust or Moon dust. Using the cutters they had brought along, they took the TV camera off the vehicle and also retrieved a little dirt scoop the vehicle carried to test the density of the soil.

"Got an extra sample for you, Houston," said Bean. "The scoop still has dirt on it!"

"Well done, troops," exulted CapCom.

They cut off the TV cable and tubing and took samples of instrument glass from the vehicle for analysis back at the Lunar Receiving Laboratory.

Since all was going so well, Houston suggested that the explorers collect dirt and rock samples at Block Crater on their way back to the LM. Conrad and Bean were both puzzled by the fact that the dust covering the Surveyor structure appeared brown, but the topsoil itself was not brown or tan at all. It was gray.

The going became a little more difficult as they kangarooed off toward Block Crater. Bean said the dust was a bit deeper and softer. He told Conrad he wanted to take a break for a few minutes. Conrad agreed.

CapCom then asked, "Pete, Al, could we have an EMU check?" The EMU is an acronym for extra-vehicular mobility unit—the Moon suit together with its liquid cooling system, portable life-support system, oxygen-purge system, and visor weighing 183 pounds on Earth and 30.5 pounds on the Moon. It was as formidable a suit of environmental armor as man has ever devised. And it worked. Conrad and Bean reported they still had 36 percent of their oxygen—plenty to get back to the LM.

Slowly, they made their way back to the Intrepid, carrying the rocks and about 25 pounds (Earth weight) of Surveyor parts. As they loped back, they came to a crater where the missile that made it seemed to have gone through bedrock. On several occasions, both explorers believed they saw outcrops of bedrock, but this turned out not to be so. What they did find and what they collected were fine-grained basaltic rocks.

Back at the LM landing site, they rolled up the aluminum-foil solar wind collector and managed to become entangled in the useless television cable, which Conrad explained was driving him crazy. It was not the first time it had snaked around their boots.

Nevertheless, both explorers were in high spirits. Conrad sang to himself. It was the first time anyone sang on the Moon.

Dust was everywhere and as they moved the rock boxes and equipment back into the LM, they spilled dust over everything. Nevertheless, they dusted off their Moon boots as they climbed up the ladder to the cabin.

At Houston, it was 1:42 A.M., CST, November 20, 1969, when the hatch of the lunar module Intrepid was sealed for the last time. Conrad then remembered he had left a magazine of color film with some photos of the Earth rising on the surface, but he had brought up to the cabin all the black and white pictures including those of the Surveyor.

During the early morning hours the two explorers cleaned up the cabin, stowed their gear and samples, and conferred with Houston. All systems were checked and rechecked. All systems were go for liftoff from Oceanus Procellarum at 142 hours, 3 minutes, and 47 seconds of the mission.

At 8:25 A.M. on the morning of November 20, the powerful LM ascent engine, with 3500 pounds of thrust, ignited and the ascent stage shot up from the descent stage, which served as its launch pad, like a shell from a cannon. Straight up it hurtled on a pillar of pale fire and then it curved over to chase the Yankee Clipper, cruising 80 nautical miles ahead.

"Away we go, Houston," sang out Conrad.

Four minutes passed and Houston called: "Intrepid, Houston. You're looking good. The harbor master has cleared you into the main channel. You're at 49,000 feet."

Flying a lower orbit than Yankee Clipper, Intrepid overtook the command ship. At 9:47 A.M., Conrad saw Clipper in the distance. The miles separating the two vehicles diminished and at 11:35 A.M., Gordon in the Yankee Clipper pointed his television camera at the Intrepid to show the little vehicle approaching at 30 feet per second.

GORDON: Hello, Intrepid. How can you look so good if you're so ugly?
Okay, Pete, you're looking good so keep on coming.

The Intrepid coasted in for rendezvous and Gordon, maneuvering the big Apollo, performed the docking.

CONRAD: Say, Houston, Intrepid. What time is it back there, anyhow?
CAPCOM: It's just about high noon. It's 9 minutes after 12.
INTREPID: Al wants to know what day it is. I've completely lost all sense of night and day.
CAPCOM: It's November 20 and it's Thursday.
INTREPID: Okay, Thursday. Roger.

Notes

1. Apollo 12 Preliminary Science Report, NASA, 1969.
2. Latharn, Gary, Press Conference, Cocoa Beach, Florida, November 13, 1969.
3. Morison, Samuel Eliot, *The European Discovery of America.* Oxford University Press, New York, 1971.
4. U.S. Geological Survey, *Geological Atlas of the Moon.* Department of the Interior, Washington, D.C., 1973.
5. Moore, Patrick, *A Survey of the Moon.* Norton, New York, 1953.
6. Paine, Thomas O., Apollo 12 Preliminary Science Report, NASA, 1969.
7. Arabian, Donald D., News Conference, Manned Spacecraft Center, November 21, 1969.

6

The Rocks of Ages

Shortly after Conrad and Bean returned to the Apollo command module, the LM, which they had cast adrift, was sent crashing to the surface by radio control of its ascent engine from Houston. The crash served a dual purpose. It provided an artificial impact test for the passive seismometers that had been deployed in Oceanus Procellarum, and it removed the LM as a hazard to navigation for future missions.

The LM hit the ground about 40 miles from the landing site. The results were stunning. All three long-period seismometers in the package recorded the impact, which set up a sequence of reverberations lasting more than an hour. Nothing like it had ever been heard on Earth. The Moon literally rang like a bell.

In fact, the seismic signal was still coming in on the recorder at Houston where Maurice Ewing, co-investigator on the experiment, told the late afternoon news conference of the astonishing event. The Moon was ringing and ringing . . . and ringing.

"Well, I must say that it was most gratifying to see this signal come in on all three components and the duration of it was certainly surprising to us—as many other things observed on the Moon have been—for instance, the glass beads," Ewing said. "As for the meaning of it, I'd rather not make an interpretation right now. But it is as though one had struck a bell, say, in the belfry of a church a single blow and found that the reverberation from it continued 30 minutes. It means either a very high Q or a very unstable structure*—so that a lot of collapses were triggered by the blow."

* Q refers to a quality factor specifying the attenuation of elastic energy in a vibrating system.

The impact occurred at 4:15 P.M., CST, November 20, 1969. The news conference began shortly after 4:30 P.M., and the Moon was still "ringing" when the seismic experimenters arrived at the news center from their laboratory.

"This only happened a few minutes ago," said Frank Press, co-experimenter. "In fact, it is still happening."

The experimenters had attempted to construct a hypothesis to explain the lunar bell effect as they hurried across the campus of the space center to meet the press.

"We have been exploring for a hypothesis walking over here and usually when we speak too soon we are wrong," Press commented. "But let me say that none of us have seen anything like this on Earth. In all of our experience, it is quite an extraordinary event. That this rather small impact . . . produced a signal which lasted 30 minutes is quite beyond the range of our experience. So, whatever it turns out to be, I think it will represent a major discovery, completely unanticipated, about the Moon."

The seismogram of the LM impact showed that impact energy built up to a peak for 7 to 8 minutes and then gradually decreased in amplitude for 55 minutes.

The seismologists looked forward impatiently to Apollo 13, when the third stage of the Saturn 5 launch vehicle would be dropped on the Moon. This huge rocket stage, 58.5 feet long and weighing about 30,050 pounds minus fuel, would produce an impact many times that of the small LM ascent stage, which was 12 feet 4 inches long and weighed about 4,760 pounds, dry. The big, third-stage rocket would really ring the lunar bell.

The ringing of the bell suggested to a number of investigators that the Moon is now a solid body with little, if any, liquid, in contrast to the Earth, where seismic signals indicate the presence of a liquid nickel–iron core. Nevertheless, it appeared from other evidence that however solid the Moon is today its present structure represented a final crystallization from an earlier molten state.[1]

Whether or not this hypothesis would hold up on further investigation, it was clear at the time of Apollo 12 that Apollo investigators had "discovered" a new kind of planetary structure in the Moon. How it evolved, as well as the details of its nature, represented one of the most elegant mysteries of the twentieth century.

Never before in all the history of science had such a vast effort been mobilized to deal with a geophysical problem—a situation that critics of NASA and its budget seized to question the wisdom of the whole lunar adventure at a time of mounting social turbulence.

The ALSEP of Apollo 12 was the most sophisticated array of re-motely-controlled instruments built up to that time. It far exceeded any remotely-controlled system of detection and measurement on the sur-face of the Earth. Its seismometer and magnetometer were incredibly sensitive and within a year provided data that revealed an astonishing, yet confusing, story of planetary development. The new data not only shed new light, but raised new questions about the cosmogony of the Solar System.

A Cause of Moonquakes

The seismometer sensor unit contained three matched, long-period instruments to measure one vertical and two horizontal components of surface motion. The sensor also had a short-period instrument sensitive to vertical motion at higher frequencies.

(The sensor consists of a mass suspended from a spring. When the ground moves, the frame to which the spring is attached also moves; the suspended mass tends to remain fixed in space because of inertia. Relative motion between the suspended mass and the frame is recorded. The recording is amplified electronically and, in the case of the Apollo 11 and 12 instruments, radioed to Earth.)

On the Moon it is possible to detect vibrations much fainter than any that could be identified on Earth against the background noise of man's activities, ocean tides, and weather. The lunar instrument was capable of magnifying vibrations 10 million times—an amount of amplification quite useless on earth with its vast background noise.

After the first year of operation, Latham reported there was no longer any doubt that seismic signals on the Moon were "utterly differ-ent" than any signals recorded on Earth.[2] In addition to the LM im-pact, a second man-made signal was recorded with the crash on April 15, 1970 of the S4B (third) stage of the Saturn 5 rocket that had launched Apollo 13 on an ill-fated lunar journey. After boosting Apollo 13 Moonward, the third-stage rocket reached the Moon on a ballistic trajectory and hit the surface 135 kilometers from the Apollo 12 land-ing site. It produced another signal that lasted a long time and died out slowly. It was clear that the structure of the mare is quite different from typical crustal structure on Earth, according to Latham.

During the year seismic signals were recorded at the rate of one a day. It was difficult to tell whether these were quakes or meteorite impacts. However, over a period of 9 months, nine matching events were recorded, all seeming to come from the same area with the same amount of energy.

Latham and his associates interpreted these as Moonquakes. It was unlikely that meteorites would hit the same area so consistently. Moreover, the signals arrived once a month at or near the time when the Moon comes closest to the Earth in its monthly orbital cycle.

This "timing" of the signals suggested that the quakes were induced by tidal strain, which was produced by the rise of gravitational attraction between Earth and Moon at the time of closest approach. Similar signals had been observed at the Apollo 11 station during its brief period of operation, but the monthly frequency was not apparent then and the interpretation of these signals remained ambiguous until the Apollo 12 signals were received. It appeared that the monthly Moonquakes triggered by the Earth were occurring at many places on the Moon.

The most active zone in the Apollo 12 region was located 120 miles southeast of the ALSEP, near a set of well-developed rilles (large cracks or ditches in the crust). Latham suggested that each month as tidal strain reached maximum these fractures slipped a little, producing the Moonquakes. On a much smaller scale the quakes seemed to be analogous to those produced in California by slippage along the San Andreas fault.

The Moon is Cool

Earth's tidal stress could not account for the entire Moonquake phenomenon, however. It could be considered only a trigger that acted on another source of strain within the Moon. It was speculated by Latham and others that perhaps the slightly ellipsoid shape of the Moon is slowly settling back into a more perfect spheroid shape as the Moon recedes from the Earth—as it appears to be doing at the rate of 3 centimeters (1.17 inches) per year. Or, perhaps the Moon is expanding slightly from the release of radiogenic heat deep within it. Whatever the source of strain is, the scientists regarded it as a new and exciting problem.

Compared with the Earth, however, the Moon is seismically inactive, Latham said, speaking for the Passive Seismic Experiment Team. The seismic evidence did not show a "hot" Moon—a body with a significant amount of molten material convecting within it, like the Earth. The Moonquakes detected were so small that an astronaut standing on the surface at their epicenter would not feel any of them.

It therefore seemed improbable to the seismologists that there could be any important sections of molten material in the Moon, except for possible, isolated pockets. Latham agreed that there may be molten material at very great depth, but its activity did not in any way approach that of the Earth, where present theory visualizes moving crustal plates causing continents to drift about like rafts and the upthrust of massive mountain chains.

But, even so, something was going on in the Moon that tantalized the seismologists. Apollo 13 had been targeted to land at the Fra Mauro formation, a region in Oceanus Procellarum of great rilles, where investigators might find the type of displacement of rocks that has occurred on Earth along the San Andreas fault of California or the Great Glen of Scotland. If it could be ascertained that blocks on one side of a rille at Fra Mauro had moved relative to blocks on the other side, the displacement would indicate that the rilles actually are rifts or fractures in the lunar crust—not simply ditches dug by lava flow or hot gases in the period of mare formation 3 to 4 billion years ago.

As far as the structure of the Apollo 12 site itself was concerned, the seismic data showed no layer of solid rock. The seismic wave velocities were much too low. It was estimated that rather loose, unconsolidated material extended down for 10 or 15 miles. If there had been a primeval layer of crustal rock crystallized from an initial period of melting, it was smashed beyond recognition by meteorite impacts over 3.3 billion years —the radiogenic age of the Oceanus Procellarum rocks. There seemed to be no crust in the Apollo 12 region, just rubble that apparently became more and more consolidated with depth.

Meteorites were still falling on the Moon. Impacts of missiles about the size of a grapefruit were being recorded at the rate of one a month, and smaller ones were recorded more often. This reduced rate of fall supported a theory that the era of great impacts had ended long ago. The Moon (and Earth) had swept up most of the debris near them in the first few billion years, but large impacts had still been possible within the last billion years. If the rate of bombardment and energy of impacts attenuated with time, it was theoretically possible to date areas of the Moon by the number of craters they exhibited—once the age of the rocks in these areas was known. Crater counting was a rough stick for age determination. Nevertheless, crater density became the basis of a number of dating estimates for lunar features. It became a rather popular numbers game for determining a lunar time scale.

A Spinning Moon

Like the seismometer, the Apollo 12 magnetometer was an extremely delicate instrument. It was also an important one for determining lunar magnetic and electrical fields, each telling its own story about the structure of the lunar interior.

Residual magnetism had been found in analysis of the Apollo 11 rocks at the Lunar Receiving Laboratory at Houston and was confirmed by further analysis at other laboratories in the United States and England. There was no general agreement on its meaning. It seemed to imply the existence at one time or another of a fairly strong magnetic field, but the origin of such a field on the Moon was a mystery. It certainly did not exist now.

The Apollo 12 magnetometer reported the existence of a lunar magnetic field of 36 gammas—about $\frac{1}{1000}$ of the Earth's magnetic field. This field appeared to be a local one, assuming the source of it was a fairly intense magnet some 200 kilometers away.

The magnetometer team, including Charles P. Sonnett and Palmer Dyal of NASA's Ames Research Laboratory at Mountain View, California, was convinced that the field was not global because it had not been sensed by another magnetometer that had been orbiting the Moon for two years aboard Explorer 35.

The small satellite had reported or, rather, confirmed the existence of a magnetic "tail"—an extension of the Earth's magnetic field on the side of the planet opposite the Sun—consisting of drawn-out geomagnetic force lines extending well beyond the orbit of the Moon. When the Moon passed through the "tail" in its orbital path around the Earth, there was a marked diminution of solar wind particles reaching the lunar surface. These were warded off by the force field of the Earth's magnetic tail. When the Moon had passed through the tail, the solar wind once more reached the lunar surface.

With the two magnetometers, one in the Ocean of Storms and the other in orbit at $\frac{1}{2}$ to 5 lunar radii (540 to 5400 miles), several important facts about the interior of the Moon could be determined. One of them was the temperature of the Moon's interior, down to the very center.

The physical principle from which this fact could be ascertained was discovered by Michael Faraday, the British electrophysicist, in 1831. Faraday found that the movement of a magnetic field across a wire would set up a current in the wire. This is the principle of electrical induction on which the transformer and dynamo are based.

How does this work on the Moon? Visualize the solar wind and the magnetic lines of force it drags out of the Sun with it to the farthest reaches of the Solar System. When the solar wind and its accompanying magnetic field wash over the Moon, eddy currents are set up in electrically conducting regions of the lunar surface and interior. These currents generate a magnetic field within the Moon.

Explorer 35 measured the incoming solar magnetic field arriving with the solar wind. The Apollo 12 magnetometer measured the strength of the field that the solar wind induces. Comparing the solar magnetic field and the induced field, it is possible to calculate a profile of electrical conductivity all the way to the center of the Moon. Once a profile of electrical conductivity is constructed, a temperature profile also can be estimated from it on the theory that conductivity rises with temperature.

Comparing the results from the two magnetometers, it was found that lunar electrical conductivity increased a million times from the surface to a depth of 200 kilometers. According to Sonnett, the probable cause of the increase was a rise in temperature with depth. He estimated the rise to be about 2° to 3° per kilometer. This would indicate a temperature of about 800° to 1000° in the deep interior of the Moon—almost but not quite hot enough for a molten core.

The induced magnetism from which these estimates were derived should not be confused with the residual or remanent magnetism of 36 gammas picked up locally by the Apollo 12 instrument. When the Earth's magnetic tail swept over the Moon, the induced field disappeared because the solar wind was shut out. However, the residual magnetism that seemed to exist in the rocks persisted.

In order for a field of 36 gammas to exist in rocks that are more than 3 billion years old, there must have been a background field of considerable force at the time the rocks crystallized. That is, such a background field must have been present between 3.2, 3.5, and 4 billion years ago, when maria formation was going on and when lavas were cooling. The reason for this is that at certain very high temperatures substances lose their magnetism. Iron, for instance, loses its magnetic property at 760°C. This is its "curie" temperature, named after the effect discovered by Pierre Curie in 1893. Thus, the rocks in the Sea of Tranquility and Oceanus Procellarum could not have retained any early magnetism when they were melted 3.2 and 3.5 billion years ago. The magnetism present to this day must have been induced by a powerful field that occurred after they cooled through their curie point. So reasoned the magnetometer experimenters.

Where did the background magnetic field come from? And where did it go? What did it portend in terms of the history of the Moon? There were a number of ideas: The 36 gamma field was induced by extremely powerful solar magnetic or electric fields, which no longer exist. The remanent field was induced by the Earth's field at a time when the

Moon was close to the Earth. Early in its career, the Moon was spinning much faster than it is now, setting up the dynamo effect that is believed to produce the Earth's field. But what about a metallic core? To produce a field by dynamo action, the Moon would have had to contain a nickel–iron core that seemed to be ruled out by its density, which is lower than the Earth's.

Nevertheless, the probability that a strong magnetic field did exist billions of years ago had to be taken into account. For members of the magnetometer team, the lunar model that solved this problem was a primordial Moon with a hot core and a rapid spin.

One of the British experimenters, S. K. Runcorn of the University of Newcastle upon Tyne, England, held that a steady field of at least 1000 gamma must have existed at the time of magnetization. He concluded that the Moon possessed its own internal field from its early history until some time later than 3.2 billion years ago.[3]

The Rocks

The Apollo 12 Lunar Sample Analysis Team made its preliminary and informal report on the evidence of the rocks and dirt at a briefing for the press at the Manned Spacecraft Center, December 12, 1969. Of first importance then, but virtually forgotten now, was the quarantine report. No sign of any microorganisms was found in any of the samples. As the Apollo 11 biological investigation had shown, the Moon was utterly lifeless—until men came.

By the December conference, 46 rocks had been examined, the largest weighing 6 pounds and the smallest less than an ounce. Two of them were breccias; the others were igneous rocks of volcanic (cooled on the surface) or plutonic (cooled below the surface) origin.

The Ocean of Storms rocks were different than those at Tranquility Base in that they contained less titanium and were younger by at least a half billion years. This was an important distinction because, as Oliver Schaeffer, a chemist of the State University of New York, Stony Brook, remarked, it showed that whatever process produced the maria rocks—volcanic or meteoric—took place over at least a half billion years and was not limited to a brief time scale.

Schaeffer and his colleagues dated the rocks in two ways: from the time they crystallized and from the time they appeared on the surface after having been excavated from below by the impacts of meteorites. The surface exposure age was estimated by measuring the rare gases and radioactive products of cosmic ray bombardment, to which surface rocks are exposed on the airless Moon.

Apollo 11 surface rocks exhibited approximate exposure ages ranging from 20 to 100 million years. Schaeffer and his colleagues dated three Apollo 12 samples showing exposure of less than 100 million years.

Rocks from both sites brought back by the astronauts were much older than these exposure ages—that is, they were formed earlier than 100 million years ago. Using the potassium–argon dating system, Schaeffer's group found that the crystalline rocks returned by Apollo 12 from Oceanus Procellarum were 2.5 to 3.2 billion years old,* whereas those brought back by Apollo 11 from the Sea of Tranquility were 3.5 to 3.7 billion years old.

The difference was significant because it showed that Oceanus Procellarum rocks—at least some of them—crystallized a billion years later than some Tranquility rocks. This plainly indicated that volcanic or plutonic-rock forming episodes continued over a period of at least a billion years.

The most common mineral found in the Procellarum rocks was pyroxene, a calcium–magnesium–iron silicate, with a low calcium content but with iron enrichment. Other minerals were ilmenite, containing iron titanate; olivine, a magnesium silicate; and plagioclase, mainly a calcium–aluminum silicate. The chemical composition of the olivine indicated that it was formed at a temperature higher than the average volcanic material has undergone on Earth.

As in the Tranquility samples, the Procellarum rocks had been formed in the absence of water, indicating that water had boiled off very early in the formation of the Moon. It was probable, indeed, that the volatile elements had boiled away from the gaseous material of the nebula even before it condensed to form the Moon. But that would not account for the heating episodes that produced the lava in the lunar seas. The maria lavas were formed 1.5 to 2 billion years after the Moon condensed from the solar nebula.

Thus, the evidence from both the Apollo 11 and 12 expeditions pointed toward internal heating episodes extending over a billion years or so. However cool and passive the Moon seemed today, it appeared to have been active in the past.

Joseph V. Smith and his colleagues at the University of Chicago reported that "we have evidence that the temperature of crystallization of these rocks is 1140°C."[4] This was higher than the 1000°C that magnetic data indicated near the center of the Moon today.

Smith suggested that the Moon might have melted completely and not

* Younger ages indicated initially by the dating method were later revised upward.

merely in its upper regions as some investigators supposed. He also hypothesized that some of the melt, possibly the last part of it, was pulled to one side by the gravitational attraction of the Earth—to the near side. There were no comparable seas of lava on the far side of the Moon.

"The easiest interpretation of much of the chemistry of the rocks will be that there has been some very extensive outgassing and sweeping away of some elements by volatilization," Smith said.

The Enigma of the Fines

One of the peculiar discoveries in the analysis of lunar material was the fact that the dirt or "fines" appeared to be considerably older than the rocks—from which it presumably was derived. So was the breccia, consisting of fine material and granules welded together by heat and pressure. There were two age groups of material at the two lunar sites man had visited. The fines and breccia appeared to be as old as the Moon was supposed to be, that is, 4.6 billion years old, but the rocks were 1.5 to 2 billion years younger. It was a nice little problem.

The lunar investigators were again confronted with a mystery growing out of an alien pattern of planetary evolution. Like the seismic ringing caused by Moonquakes or impacts, the dual ages of the rocks and the fines could be explained only in terms of the Moon's particular evolution. Conversely, they were clues to the possible processes of lunar evolution.

Although most of the Apollo 11 and 12 rocks exhibited radiogenic ages of 3.2 to 3.7 billion years, there were a number of exceptions— rocks that seemed to be younger than 3 billion years and older than 3.7 billion years.* But, because of the uncertainty of the dating techniques, ages of rocks that appeared to be 4 to 4.4 and even 4.6 billion years old were regarded with some suspicion of error.

The exception, however, was a 3-ounce rock sample (No. 12013), the size of a small lemon, that Conrad and Bean had brought back from the Sea of Tranquility. It was dated as having crystallized 4.6 billion years ago—close to the assumed 4.6 to 4.7 billion year age of the Solar System. It appeared to be the same age as meteorites from which the age of the formation of the Solar System is derived.

The lemon-sized rock was dated by the rubidium-87–strontium-87 method by Gerald J. Wasserburg and his associates at the California Institute of Technology. It was quickly dubbed the "genesis" rock.

_____

* Later measurements ruled out ages of less than 3.1 billion years.

Chemical analysis showed that it was distinguished from other lunar samples by containing 20 times more uranium, thorium, and potassium.

In the light of these rock ages and those of the fines and of the breccia, it seemed reasonable to conclude that the processes of chemical differentiation and crystallization had been going on for more than ½ to 1 billion years on the Moon. Presumably they had stopped about 3 billion years ago—at least from determinations based on limited investigation at only two places.

From the data and the speculation the data generated, there was no doubt that Apollo 11 and 12 opened a Pandora's box of scientific problems. And these were the grist for the first NASA Lunar Science Conference that filled the Albert Thomas Convention Hall in Houston, January 5 to 8, 1970.

The Rock Festival

Attracting 600 scientists as participants and several hundred more as observers, the Lunar Science Conference was destined to become an annual event. Later, it would be known to its habitués as the "Lunar Rock Festival," a whimsical, rather than derogatory, reference.

At this first meeting, 144 papers were presented on the subjects of lunar age determination, trace elements, abundances of major elements and stable isotopes, rare gasses, the solar wind, mineralogy, fine particles, shock effects, magnetic and electrical properties, Mössbauer effect spectrometry,* and organic chemistry. If anyone expected that this wealth of data would resolve outstanding questions about the origin and evolution of the Moon, he faced disillusionment. As Harold Urey had remarked in the days of Project Ranger, each succeeding layer of new data could be interpreted in ways that would support conflicting theories about the Moon. The mass of evidence and speculation presented at the 1970 conference gave further proof of "Urey's law."

Since little time had been available for detailed studies of Apollo 12 samples, most of the rock analysis dealt with the samples from the Sea of Tranquility. Three new minerals were reported: pyromanganite, ferro-

---

* Mössbauer spectrometry is a technique of identifying and characterizing certain elements, especially iron. It is based on the Mössbauer effect discovered in 1958 by Rudolph L. Mössbauer, a German physicist who was then a graduate student in Munich. The effect refers to the recoil-free emission and resonant absorption of gamma radiation in certain radioactive nuclei. The effect has a number of applications. Spectometry based on it was used to identify iron-bearing minerals in lunar rocks and to determine the phase distribution and oxidation states of the iron—all important clues to lunar evolutionary processes.

pseudobrookite, and chromium titanium spinel. None of them had been seen on Earth—which did not mean they were not there, but that they had not been found. However, it could mean they were peculiar to the Moon—products of the Moon's particular chemical history that was characterized by absence of water and paucity of oxygen.

The Conference Summary drafted by a group of scientists comprising the Lunar Sample Analysis Planning Team suggested a tentative conclusion that the Moon was certainly depleted in a number of volatile elements (oxygen, potassium, rubidium, cesium, chlorine, thalium, and sodium) compared to the abundances of those elements in Earth rocks.[5] At least, the preliminary analyses of the rocks from the two sites gave evidence of the depletion. If this turned out to be the case for the whole Moon, the summary said, "one can infer that lunar material separated from a high-temperature dispersed nebula at 1000°C, or higher."

Except for "a single exotic rock fragment yielding an age of 4.4 billion years" (called "Luny Rock I"), the summary noted, the basaltic crystalline rocks at Tranquility formed 3.7 billion years ago.* "The relatively young age of the basalts shows that the Moon has not been a completely dead planet from its formation, but has undergone significant differentiation, at least locally, in a thin lunar crust."

This estimate of a thin crust turned out to be a wrong guess. The error could be attributed in part to the tendency of many investigators to think of the Moon in terms of a terrestrial framework. But the team that wrote the summary concluded that a primitive crust had been destroyed on the near side by bombardment, that the few very ancient rocks found were possibly remnants of it, and that the lunar highlands or terrae, which medieval observers thought of as continents, would reveal rocks of even greater age. In the highlands, perhaps the original crustal rocks would be found, and they would tell what a planet was like at its genesis. The primitive crust of the Earth had long ago been obliterated.

However, the team added optimistically, the surface sampled so far "gives us a picture of the early evolutionary processes of a terrestrial planet which have hitherto been obscured from view." It was the team's consensus that the natural remanent magnetism found in the rocks suggested that the Moon had a magnetic field at one time with a strength possibly of a few percent of the Earth's field. How this field was generated was still speculative.

---

* The summary did not mention three reports of rocks dated at 4 billion years and older.

The formal presentation of papers at this historic rock conference was amplified by panel discussions and news conferences. At these, particularly at the news conferences, the investigators felt freer to unburden themselves of their special views and theories than in their formal presentations—in which they felt constrained to report only their observations and experimental results.

These informal discussions reveal much more about the state of knowledge and theory concerning the Moon, the Earth, and the Solar System than the papers do, circa 1970. Predominant concepts were then beginning to emerge, particularly about postaccretion evolution of the Moon. Schools of thinking were forming. They were to play a critical role in determining where future expeditions would go and what the explorers would be required to do.

The predominant view was that the Moon had formed "hot" and that water, oxygen, and other volatiles were lost during the condensation process. Melting near the surface came a billion years later and persisted, perhaps intermittently, for another $\frac{1}{2}$ to 1 billion years. The source of heat was radioactive decay. About 2.5 to 3.2 billion years ago, the Moon had cooled off sufficiently to allow it to subside into the relatively inactive planet we find today.

The Rings of Earth

Whether the Moon had formed hot and was now cold; whether it had formed cold, heated up, and then cooled off long ago; or whether it had formed hot and had experienced intermittent periods of heating until geologically recent time could not be determined from the Apollo 11 and 12 data alone. But a start had been made. There was significant dissent from the hot–cold model, which seemed to be emerging as the predominant view of the first lunar science conference. Urey still held out for cold accretion but admitted that internal heating from radioactivity was the best way to account for the basaltic rocks that were younger than the supposed age of the Moon.

As expected in such a controversial conclave, the scientists could not pass up an opportunity to elaborate on their theories of lunar origin.

John O'Keefe of the Goddard Space Flight Center persisted in his Earth fission theory of lunar origin, finding new support in the Apollo 11 and 12 data—as per Urey's law. He based much of his argument on the low abundance of siderophile (iron and iron-loving) elements in the lunar rocks compared to their abundance in meteorites. The iron and nickel that was missing on the Moon (and also relatively low in the

Earth's crust) had gone into the core of the Earth before the mass that became the Moon separated from the mass that became the Earth, O'Keefe maintained.

Several geochemists, notably Paul Gast then associated with the Lamont-Doherty Observatory, insisted that the Moon could not have been made "by pulling it off the Earth" because of differences in chemical isotopes between lunar and Earth rocks. Lunar rocks appeared to be more depleted in the element rubidium. However, Gast qualified his opinion to concede that if fission did happen, it would have had to occur during the first 200 to 300 million years of the Moon's history.

This view of the fission theory was adopted by the Lunar Sample Analysis Planning Team, which stated in its summary: ". . . if the Moon formed from the Earth, it can now be stated with some confidence that this separation took place prior to 4.3 billion years ago."

However, comparison of element abundances in the Earth and the Moon could be quite ambiguous. John A. Philpotts and C. C. Schnetzler of the Goddard Space Flight Center reported that although potassium, rubidium, and strontium were considerably lower in eight Apollo 11 samples than in the Earth's crust, the lunar sample trace element concentrations were similar to those expected for the bulk Earth. They suggested that "this might tend to favor a fission hypothesis or a propinquitous aggregation hypothesis, rather than the capture hypothesis for the origin of the Moon." On the other hand, they said, similar trace element characteristics of the Earth and Moon could also indicate that similar processes of condensation were going on in different parts of the solar nebula.[6]

Urey (and most others) believed that the Moon had been formed separately from the Earth, then captured. No one had come up with an acceptable theory of how capture had come about. The depletion of volatile elements raised the possibility that the Moon was formed in a hotter part of the solar nebula than the part in which the Earth condensed. But the depletion also could have occurred as the result of very early cooking after the Moon had condensed. Nothing was proved.

A. E. Ringwood, a geophysicist of the Australian National University, took the same preliminary data that O'Keefe used to support the fission theory and applied it as proof that the Moon could not have fissioned from the Earth. The difference in chemical element abundances in lunar and Earth basalts are inconsistent with the fission hypothesis, said Ringwood. If the Moon had fissioned from the Earth, he maintained, the basalts in each would show closer chemical compositions.[7]

The only origin theory that could explain the differences in the basalts was the Saturnian-ring theory of Öpik (also Gilbert), according to Ringwood.

As I have related earlier, the ring theory holds that the Moon coagulated from a sediment ring of planetesimals orbiting the Earth. These materials were analogous to the rings of Saturn, but the total ring was much more massive. It was derived, Ringwood hypothesized, from a primitive atmosphere that formed during the later phases of the Earth's accretion. During accretion, he speculated, the temperature was high enough to vaporize silicates as they were accreting on the Earth. When the silicate vapor cooled, the silicates were precipitated into planetesimals—balls of rock, analogous to hailstones forming in the clouds of our present atmosphere. These began to coalesce as they orbited the Earth and eventually formed the Moon.

As early as 1961, Öpik proposed that the "observed small systematic deformations" of pre-mare craters "are only compatible with the Moon having accreted from a cloud of fragments circling the Earth in direct orbits at a distance of 5 to 8 Earth radii (20,000 to 32,000 miles) or greater."[8]

A study of random distribution of low-angle impact craters on the lunar surface "rules out the possibility that particles initially responsible for the origin of such craters had, prior to impact, been in heliocentric orbits," according to Zdenek Kopal, of the University of Manchester, England.[9] "The observed facts are more consistent with a view that particles responsible for most of the large primary impacts at the earliest stages of lunar history were moving with the Earth–Moon gravitational dipole and may have represented leftovers from the formation of this pair of cosmic bodies."

Kopal drew no conclusion about lunar origin from this indication that the objects bombarding the Moon early on were in orbit about the Earth–Moon system, except to remark: "This would, in turn, imply that the Earth and Moon were already gravitational partners (though not necessarily at their present distance) during the first few hundred million years of their existence—an epoch in which the cratering of the lunar surface is now thought to have been largely accomplished. And if the Moon was captured by the Earth, this event must have occurred prior to the main period of cratering of the lunar continental land masses."

The Core of the Matter

Closely related to the controversy about the origin of the Moon was the argument about whether it had a core. Urey had insisted that the

Moon could not have a core. As previously noted, its average density was 3.3 times that of water whereas the average density of the Earth is 5.5 times that of water (which is 1 gram per cubic centimeter).

The average density of the Moon, as I have noted, appears to be the same as that of the Earth's mantle, the region lying just below the crust. Since the Earth's crust exhibits an average density of only 2.7, it stands to reason that something very large and very dense indeed must exist deep inside the Earth to raise its average density to 5.5, namely a nickel–iron core, so hot as to be molten. The existence of such a core has been confirmed for many years by earthquake seismology and no serious group of scientists questions it.

Thus, Urey and others insisted that the Moon could not have a core because of its lower average density, but some geophysicists were not convinced. Lunar soils at the Apollo 11 and 12 sites exhibited density of 3.1. They were denser than Earth soils, but less dense than the whole Moon average of 3.34 grams per cubic centimeter. There was something a little denser in the Moon than the topsoil, it appeared. Could there be a small core?

But until the remanent magnetism of the lunar rocks was found, the likelihood of even a small core appeared to be dim. Further, it has been observed that the Moon's moment of inertia is close to that of a homogenous sphere.

However, on the basis of Apollo 11 and 12 magnetic findings, the probability of a core began to grow, even though it could not be a very dense core. Edward Anders, a meteorite expert from the University of Chicago, remarked during a panel discussion at the Lunar Science Conference that a core of "something like 30 percent of the Moon's radius (324) miles) could not be excluded." Anders noted that Lunar Orbiter photos of the floor of the Crater Tycho showed formations of rope lava.

Inasmuch as Tycho is one of the youngest of the big craters, with an estimated age of 800 million years or so on the basis of crater counts,* the photographic evidence of lava there implied that the Moon has been hot and active during most of its history—not merely for the first billion years of it. But the photos could not indicate whether the lava was volcanic or the product of impact heating.

Anders' remarks drew an impatient reply from Urey, who reiterated his arguments for a cold, rigid, and homogenous Moon and added: "I

* Many investigators, starting with Gilbert in the last century, have pointed out that the density of craters in the lunar seas corresponds closely to a declining rate of asteroidal and cometary impacts over time. The number and size of the impacting objects declined with time as these objects were "swept up." Crater density is noticeably higher in the older formations and thus is used for rough dating.

am a little shocked to find that of papers presented here today from Chicago, of all places, where I started my arguments in regard to this— and wrote a book on this subject, gave lectures on this subject—that these papers ignore completely these arguments of 20 years. And also that a colleague from Chicago repeats them [hot Moon arguments] from this table."

Anders had reported that only 2 percent of the lunar soil he examined came from meteorites, mostly the type known as carbonaceous chondrites. These supply 98 percent of the gold on the surface, he told a news conference . . . hardly enough gold to finance space programs.

The important fact in the soil and rock composition, however, was the low abundance of volatile elements. This convinced most observers that very early in its history the Moon went through a high-temperature stage when these elements were cooked off. For example, Anders said, nearly all the primeval lead boiled away, along with much of the potassium, sodium, rubidium, and cesium. Most of the lead found in the samples was radiogenic, Anders said, lead that is formed from radioactive decay of uranium and thorium.

The depletion of the volatile elements became a pivotal point, not only for the evolution of the Moon but for the grasp of planetary formation in general. Paul Gast commented during the conference: "My picture is that this may have happened before the material that we now see on the Moon actually was accreted to form the Moon—that this happened at a very high temperature stage, very early in the history of the Solar System, when the solids were being separated from the gaseous material."

Gast said he supposed that the lunar exterior was melted to a depth of at least 100 kilometers by impact or internal (radioactive) heat some 3.65 billion years ago when the Tranquility lavas were generated. Impact heating was a difficult idea for some people to accept because it implied a bombardment of incredible intensity. It raised the question of where the meteorite or asteroid missiles had been stored during the billion years between the supposed time of lunar formation 4.6 billion years ago and the time Mare Tranquillitatis lavas cooled 3.5 to 3.7 billion years ago.

Some investigators, such as John A. Wood of the Smithsonian Astrophysical Observatory, ruled out the decay of long-lived radioactive isotopes as the main source of heat for melting rocks, but conceded that short-lived radioactive isotope heating might account for later lavas. He contended that the heat from long-lived radioisotope decay would have been dissipated before it could have accumulated to reach melting temperature. Wood supposed a hot, early Moon that was heated by impact

energy and by the retention of accretion and preaccretion high temperatures to account for the chemistry of the Apollo 11 and 12 samples.

The Ancient *"E"*

The question of why the lunar fines and the breccia appeared to be a
billion years older than the rocks they presumably came from persisted
throughout the conference. In vain did one expert after another attempt
explanations that (1) the fines and breccias contained pieces that
(somehow) had not been recycled from more primitive material and
thus retained a radiological signature of very early crystallization and
(2) the lunar soil was a conglomeration of materials of all ages of
volcanic or plutonic activity, and in this conglomeration were bound to
be primeval pieces, such as rock fragments more than 4 billion years
old.

Of all the explanations for this anomalous situation, perhaps those
advanced by Wasserburg and Anders were the most coherent. The two
chemists shared a similar intellectual approach to the questions about
the origin and evolution of the Moon, were of established academic
stature, and were both highly regarded. There the similarity ended. Two
more unlike personalities would be hard to find in the same academic
discipline.

Gerry Wasserburg was volatile, colloquial, seemingly impulsive. He
was of medium stature, rather stocky. He moved quickly, usually dart-
ing about at a half run. His intensity was camouflaged by a flippancy,
in which wit and disregard for academic loftiness were conspicuous. He dressed casually, wise-cracked uninhibitedly, and illuminated
his explanations with puns and colorful analogies making him one of
the most quotable figures at the conference. The paper he gave on
analysis of Apollo 11 samples listed his address and that of 11 colleagues as "The Lunatic Asylum of the Charles Arms Laboratory of
Geological Sciences, California Institute of Technology," in which all
12 collaborators were designated as "inmates."

Edward Anders was tall, lean, and reserved. With occasional flashes
of wry humor Anders approached his subject with a straightforward
intensity free of digression and circumlocution. His manner was deliberate, careful, and courteously forceful. He dressed conservatively and
invariably appeared poised, but preoccupied. Yet, Anders, too, exhibited the inner excitement about his work that was so characteristic of
Wasserburg. Fanatically opposed to tobacco in any form, especially
cigarettes, Anders carried on a personal crusade against smoking and

usually displayed a lapel button denouncing it. Wasserburg exhibited no such inhibition. They were two of the stars of the Apollo lunar investigation and certainly among the greatest scientific "detectives" probing the mystery of the Moon. And both shared the rare facility of being able to translate abstruse scientific concepts into terms comprehensible to the layman.

On one occasion Wasserburg fielded a press question asking why rocks 4.6 billion years old aren't found on the Earth. The question obviously was from a newcomer to the Moon story, because journalists who had been covering the program for some time already knew the answer. But Wasserburg always answered a question, even if he thought it was naive or showed lack of preparation. He explained:

> Well, the oldest known rocks on Earth are 3.4 billion years old. The largest areas of old rock on Earth are 2.65 billion years old. There are very few 3.4 billion years old, so the first thing is you have to understand the Earth is a pretty lively planet. Namely, the time for something to get remade on the Earth is . . . a half billion years by remelting and erosional processes. So it's very hard to find anything that's old. It's sort of like how many 300-year-old men do you know? So they die and they get reborn into cockroaches, Archie and Mehitabel, sometimes cats, but there is a rejuvenation process by remelting which gives you an individual age. . . ."[10]

Anders approached the question of difference in ages between the rocks and fines on the Moon in this way:

> Let's take a hunk of iron. We seem to agree that any piece of iron that you pick up on Earth is likely to be older than 4.5 billion years because it was made in the interior of some star (before the formation of the Solar System).* And then General Motors gets hold of it and makes it into a car. This car has a 1970 nameplate; it is clearly a 1970 model. It was put through a great deal of processing and converted into a car.
>
> The nameplate assures you that a great deal has happened to this shapeless hunk of metal in the year 1970. Then a few years later the

---

* He referred to the theoretical process by which the elements are synthesized in stars, starting with the transmutation of hydrogen to helium by nuclear fusion, the conversion of isotopes of helium to carbon and oxygen isotopes and their transmutation into neon, magnesium, silicon, and sulfur isotopes. According to Fowler, as successive nuclear processes continue the composition of a star changes. Instabilities arise, expelling the transmuted material to space where it is mixed with uncondensed hydrogen gas in the galaxy and is available for condensation into second and later generation stars.[11]

car is wrecked. It is converted into scrap metal. The evidence of the 1970 "event" has been erased and once again this is a 5 billion year old piece of iron.

This is the situation of the lunar rock. The Tranquility Base rocks have gone through a process that changed their chemistry 3.7 billion years ago and sorted them out into several different kinds of rocks which we have called *A* and *B*. There has got to be an *E* somewhere— we conclude that.* Then if you make the soil by mixing *A* and *B* and *E* by impact, spice it with 4 percent highland material and 2 percent meteoritic material and then analyze the whole bulk sample, you once again establish the fact that this material was made 4.5 billion years ago, because you really destroyed all the careful, chemical separation that took place 3.7 billion years ago.

On Earth, it was reiterated, the first billion years of material was lost. It was pulverized by meteoritic bombardment, eroded by winds and water, and cooked and recooked over and over. The process of erosion, deposition, sedimentation, and consolidation still goes on. In this way, the Earth's primitive crust was obliterated. The only hope of finding out how a terrestrial planet, such as the Earth, Mars, Venus, Mercury, or the Moon, began its evolution was to find rocks and soils that would fill in the gap of the first 1.5 billion years missing on Earth. The Moon was one laboratory where this evidence might be found. Mars possibly was another. But now we had arrived at the Moon, and Mars was perhaps a half century away in terms of manned exploration.

One of the surprising pieces of evidence was the Apollo 12 "genesis" rock, 12013. Chemically, it resembled a piece of Earth granite, one of the most chemically fractionated (cooked) products of the crust. Its great age suggested two possibilities: that its chemical processing had taken place in the first 500 million years after accretion, which meant that postaccretion heating had started earlier than many experts had suspected, or that the Moon had been formed cold and its outer surface had become hot in its first half-billion years and had bubbled and frothed from radioactive heating. Instead of being boiled off in the solar nebula, volatile elements and water could have been vaporized to form a primitive atmosphere, which was blown away by the solar wind or dissipated into space by thermal energy exceeding the low gravitational force of the Moon.

---

* The letters, *A* and *B*, were used by the preliminary examination team to designate fine-grained and coarse-grained crystalline rock. The letter *C* referred to breccia and *D* to the soil. Thus, in Anders' nomenclature, *E* referred to a missing ingredient that he believed had to be present to account for the age of the soil.

When he was asked which hypothesis he liked, Wasserburg recited a little verse:

> Some like it hot,
> Some like it cold.
> Some like it in the pot,
> 4.6 billion years old.

There the matter rested, for the time being.

The question of whether lunar material had undergone its initial fractionation in the solar nebula before condensing into the Moon or after condensation as a result of radioactive heating in the Moon could hardly be resolved on the basis of only two lunar landing missions. A wider range of samples was needed and the search for a remnant of the primeval crust led to the Fra Mauro formation. This is a hilly region about 110 miles east of the Apollo 12 site covering a large area around Mare Imbrium (Sea of Rains), a depression some 700 miles in diameter thought to be an impact structure—the largest circular one on the Moon.

The Fra Mauro formation, named for a fifteenth-century Italian monk and cartographer, consists largely of a blanket of material ejected from the Imbrium impact. On this blanket, explorers might find rocks belonging to the original lunar crust. Perhaps these samples would tell whether the primeval Moon had a crust already cooked or one that was heated up and fractionated later.

The next port of call, therefore, was Fra Mauro. A valley in this rocky region was the target for Apollo 13.

The Oldest Rocks

Meanwhile, the mysterious E component that Anders had mentioned had become evident to several investigators. Not only was the soil older than the rocks, but the E component was older than the soil. Could it be a relict of the primitive crust?

By measuring the disintegration of uranium and of thorium to lead, several investigators had dated Apollo rock fragments as 4 billion years old and older and dust and breccia 4.6 to 4.63 billion years old—as old as the Solar System.

Leon T. Silver, of the California Institute of Technology, reported that four individual rocks that he dated by the uranium–lead method yielded ages of 4.13 to 4.22 billion years, whereas the dust and breccia gave ages of 4.6 to 4.66 billion years.[12] Mitsunobu Tatsumoto and John N. Rosholt, of the U.S. Geological Survey, Denver, reported ages of 3.78, 3.87, and 4.07 billion years for crystalline rocks and 4.66 billion years for the dust and breccia.[13] In addition, rubidium–strontium measurements of six rocks by V. Rama Murthy of the University of Minnesota and R. A. Schmitt and P. Rey of Oregon State University indicated ages of 4.42 billion years plus or minus 240 million years.[14]

Silver noted that the four rocks, old as they appeared to be, were still 400 million years younger than the dust and breccia. "If the four rocks represent sources which have contributed to the dust also, and many data suggest this, then it may be inferred that the dust contains some materials with higher lead-207–lead-206 ratios than measured from the composite sample.* This would imply an even greater age than 4.63 billion years for some parts of the lunar surface."

The mysterious *E* became a missing link in the history of the Moon. Without such a component, Silver said he found it difficult to reconcile the significant (age) differences between the rocks and the regolith (dust and rubble). It seemed to him that the age discrepancy suggested fundamental and unexpected rock-forming processes on the Moon of great antiquity. However the Earth acquired the Moon, by capture or fission, the earliest rock-forming process must have taken place before the Moon had fully accreted—or so the record of the very oldest materials in the rocks and fines indicated.

Two investigators, Patrick M. Hurley and William H. Pinson, Jr. of the Massachusetts Institute of Technology, found differences in the amount of rubidium in two groups of lunar rocks.[15] The differences suggested to the MIT scientists that the Moon was made up from an assortment of materials that had condensed in different parts of the solar nebula.

Although the dating game tended to mystify rather than clarify the genesis of the Moon, a predominant view crystallized at the first Lunar Science Conference about the age of things on the Moon. It was tentatively agreed that the evidence showed that the Sea of Tranquility had formed 3.6 to 3.7 billion years ago and, as mentioned earlier, that the Ocean of Storms, at least at the Apollo 12 site, was ½ to 1 billion years younger. Later the difference would be narrowed when rock ages were recalculated.

The Sea of Tranquility dates were based largely on rubidium–

---

* Uranium-238 and uranium-235 decay to lead-206 and lead-207, respectively, whereas thorium-232 decays to lead-208. These lead isotopes are regarded as "radiogenic," that is, the product of radioactive decay. The age of a rock can be determined by measuring the ratio of these lead isotopes and the remaining uranium and thorium in the sample.

strontium measurements of five rocks by Wasserburg and his fellow "lunatic asylum inmates" at the California Institute of Technology[16] and on potassium–argon measurements of seven rocks by Grenville Turner, a physicist, and his associates at the University of Sheffield, England.[17] They were confirmed by other measurements, including those of Paul Gast and Norman J. Hubbard of six igneous rocks by the rubidium–strontium method.[18]

Beyond the age of the maria loomed the much greater age of the Moon itself from the measurements of Silver, Tatsumoto, and others, and this seemed to be greater than the generally accepted 4.6 billion years.

The "genesis" rock was a fascinating anomaly. Where had it come from? The best guess was that it had been heaved out of Copernicus. About 850 million years ago, the rock had undergone measurable shock and high temperature and pressure. These effects, determined at the "lumatic asylum," suggested the age of Copernicus.

Neither Tatsumoto nor Silver found their measurements of 4-billion-year-old rocks and 4.6-billion-year-old dust inconsistent with the 3.6-billion-year-old estimate for Sea of Tranquility formation. During conference discussion, Tatsumoto suggested that the dust and breccia he dated at 4.6 billion years could have been formed at an earlier stage of lunar history than the crystalline rocks of 3.5 billion years of age. Silver said that unquestionably he and Tatsumoto were seeing evidence of the missing ingredient—the mysterious E.

But the indications of rock ages older than 4.6 billion years were greeted with skepticism, for that was the accepted age of the Solar System.

Back to Fundamentals

One of the most articulate investigators at the first conference was Gast. He had a rare knack of arranging divergent arrays of data into a coherent focus. His death from cancer in 1973 was a great loss to science in general and to the lunar program in particular. At the final news conference of the Houston meeting, he summarized the basic question that had emerged so clearly from the mass of data and speculation as follows:

Now the question is, where did this fractionation take place? Did it take place at some early stage in the history of the Moon when the Moon was so hot that these materials—these volatile elements, including water—boiled off and were vaporized into space, or was the material that went to form the Moon from the various, primitive materials in the solar nebula already different from this chondritic-meteorite parent material* and was the Moon accumulated out of some primitive material that was depleted in these volatiles?

That's a very important question and that's something which we had some speculation about with regard to the Earth. The Earth is also depleted of some of these elements, not as much as the Moon. Now we're back to the very fundamental question about how did planets originate—how did they form—what were the chemical conditions that existed at a time when planets were formed, perhaps at this very early stage that we heard about at the banquet when the Sun was a totally different star from what it is today, a much bigger star, and the temperatures at the distance of the Earth were much hotter than they are to-day. The fact that we're talking about these kinds of questions, I think, says, yes, we are beginning to answer the question of the origin of the planets.

The banquet reference was to a dinner address during the conference by the British astrophysicist and astronomer, Fred Hoyle, who had admitted that a lot of the geophysical talk was out of his field. Hoyle remarked:

> I'm at a loss to understand all of the funny, mineralogical names. In fact, I'm at a loss to understand just about everything that's being talked about.
>
> As an astronomer, I naturally find myself approaching problems connected with the Earth and Moon with the idea that some 4.5 billion years ago the abundance of material separated from the Sun. At that time, the Sun was very likely much more luminous and much more ascended in size than it is at present. Gradually, the priority gases cooled, possibly because they were pushed farther away from the Sun and because the Sun became less luminous as time went on. As the temperature of the gases declined, solids and liquids condensed with the planetary material. These first condensations were chemically differentiated in accordance with the general binding of moleclues and with the way the molecules were themselves bound into various solids and liquids. So the beginning was characterized by a quite high degree of differentiation.

Hoyle said it could be demonstrated that the general temperature at which the materials composing the Earth and Moon condensed was about 1500° absolute—which is also the temperature at which rock melts to form magma or lava. "I therefore believe myself that the first

---

* Chondritic meteorites are assumed to represent the original elemental composition of the solar nebula and have the same element abundances as the Sun.

considerable bodies to be formed in the Solar System were already highly differentiated because they were formed from individual pieces that differed significantly from each other," he said.

Notes

1. Anderson, Jr., A. T., Crewe, A. V., Goldsmith, J. R., Moore, P. B., Newton, J. C., Olsen, E. J., Smith, J. V., and Wyllie, P. J., "Petrologic History of the Moon Suggested by Petrography, Mineralogy and Crystallography," Lunar Science Conference, Houston, 1970.
2. NASA News Conference, Washington, D.C., November 18, 1970.
3. Runcorn, S. K., "Implications of Magnetism and the Figure of the Moon;" Third Lunar Science Conference, Manned Spacecraft Center, 1972.
4. Smith, J. V., et al., Lunar Science Conference general discussion, Houston, January 5, 1970.
5. Apollo 11 Lunar Science Conference Summary, *Science*, January 30, 1970, Vol. 167, No. 3918.
6. Philpotts, J. A., and Schnetzler, C. C., "Potassium, Rubidium, Strontium, Barium and Rare Earth Concentrations in Lunar Rocks and Separated Phases," Lunar Science Conference, Houston, 1970.
7. Ringwood, A. E., and Essene, E., "Petrogenesis of Lunar Basalts and the Internal Constitution and Origin of the Moon," Lunar Science Conference, Houston, 1970.
8. Öpik, Ernst J., "Tidal Deformation and Origin of the Moon," *Astronomical Journal*, Vol. 66, No. 2, March 1961.
9. Kopal, Zdenek, "Cosmic Influences on the Early History of the Lunar Surface," a paper presented at the NATO Advanced Study Institute on Lunar Studies, Patras, Greece, September 1971.
10. Wasserburg, G. J., Lunar Science Conference News Conference, January 18, 1970.
11. Fowler, William, A., "Origin of the Elements," in *The Scientific Endeavor*, a report of the Centennial Celebration, National Academy of Sciences, 1963.
12. Silver, L. T., "Uranium-Thorium-Lead Isotope Relations in Lunar Materials," Lunar Science Conference, Houston, 1970.
13. Tatsumoto, M., and Rosholt, J. N., Age of the Moon: An Isotopic Study of Uranium-Thorium-Lead Systematics of Lunar Samples," Lunar Science Conference, Houston, 1970.
14. Murthy, V. R., Schmitt, R. A., and Rey, P., "Rubidium-Strontium Age and Isotopic Abundances of some Trace Elements in Lunar Samples," Lunar Science Conference, Houston, 1970.
15. Hurley, P. M., and Pinson, Jr., W. H., "Rubidium-Strontium Relations in Tranquility Base Samples," Lunar Science Conference, Houston, 1970.
16. Wasserburg, G. J., et al., "Ages, Irradiation History and Chemical Com-

position of Lunar Rocks from the Sea of Tranquility," Lunar Science Conference, Houston, 1970.

17. Turner, G., et al., "Argon 40–Argon 39 Dating of Lunar Rock Samples," Lunar Science Conference, Houston, 1970.

18. Gast, P. W., and Hubbard, N. J., "Abundance of Alkali Metals, Alkaline and Rare Earths in Lunar Samples," Lunar Science Conference, Houston, 1970.

7

Odyssey and Aquarius

Hopes of the Apollo lunar scientists for resolving major questions about the origin and evolution of the Moon reached a peak of optimism at the beginning of 1970. Apollo 11 and 12 had demonstrated that men were able to function as explorers in the lunar environment, the harshest man had penetrated. Even though such duties as collecting rocks and dirt and setting up instruments seemed menial, they were viewed by the advocates of manned space flight as ample justification for the enormous investment they required.

Scientifically, the climax of the early Apollo program was the first Lunar Science Conference in Houston. Two successful missions had produced a volume of data that would take years to analyze fully. The papers offered at Houston presented only preliminary results. Ten lunar landings had been scheduled by NASA in 1969. Two were accomplished; there were eight more to go. Then the first of a series of blows that was to beat this program to its knees was struck. The day after the Lunar Science Conference opened, George M. Low, NASA deputy administrator, announced that budget limitations had forced the agency to cancel Apollo 20, the tenth lunar landing mission. He explained that the Saturn 5–Apollo earmarked for this mission would be used instead to launch a space station.

Reaction to this deletion in the lunar program was subdued among the scientists and observers at the conference. It appeared as though NASA had compromised with budget critics and made a trade-off—a manned obital laboratory for a Moon mission.

The cancellation of one flight still left a program of nine landings, now seven more to go. The optimistic outlook of the Apollo scientific community, a worldwide one, remained undimmed. However, some scientists sensed that the lunar program was being increasingly isolated from the main stream of national interest and attention. Even though international interest remained high, Apollo no longer dominated the world stage in 1970. It was receding at home as a primary focus of American enterprise and ingenuity.

The international chariot race to the Moon that grew out of missile competition with the Soviet Union and launched Apollo ended when the Eagle landed. With astonishing rapidity, the reason for being of the Apollo program had undergone a metamorphosis. Overnight, it became a scientific undertaking of the highest intellectual purpose. The horse race had finished as a quest to probe the origin and evolution of the Moon and, through it, of the Solar System.

By the spring of 1970, it was obvious that the intellectual rationale for Apollo could not support the full program in the absence of enthusiastic public support—and that was waning. The voice of Apollo's critics began to swell in volume. It became amusing to ridicule the program in intellectual and political circles. The largest program of environmental exploration in history was being characterized by congressional and press critics as a "Moondoggle."

The support of the established scientific community began to wither. From a practical standpoint, only a relative handful of Earth scientists was benefiting from Apollo to pursue research of questionable social relevance at enormous expense—to the deprivation not only of more immediate and pressing social needs but also of more "relevant" scientific goals.

This attitude spread through many sectors of American society that had supported space research enthusiastically a few years earlier—notably academia. It marked the end of the scientific revolution that had characterized the post-World War II years, the 1950s and the 1960s. The counter-scientific revolution that broke out in 1970 quickly found support in the liberal movement. Liberals equated the funding of technological development and fundamental scientific inquiry (except in obvious medical applications) with indifference to social welfare. The same attitude was adopted by the military establishment and its committees in Congress. Generous military funding of basic research became a thing of the past under a Senate Armed Services Committee dictum that said, in effect: "If the troops can't use it, forget it." A substantial portion of the radio astronomy program in the United States that the Department of Defense had supported for years promptly collapsed. The troops couldn't use it.

With this change in attitude, the physical scientists who had been enormously influential in the shaping of science and defense policies since World War II began to fall from grace. Finally, in 1973, President Nixon eased them out of the White House, handing over the science advisory function to the National Science Foundation director, H. Guyford Stever, who reported to a presidential assistant.

This whole process of rejecting science, a form of anti-intellectualism or "know-nothingism" that this country has experienced before, was focused initially on Apollo, the most vulnerable big science project of the period—with the supersonic transport running a close second. The spearhead of the attack on big science was a group of liberal Democrats in Congress who denounced it as antithetical to a humanistic value system and its priorities. Apollo, which had been born in the Democratic Administration of John F. Kennedy in 1960, had virtually been disowned by the liberal wing of the Democratic party by 1970. It now depended for survival on the conservative Republican Administration of Richard M. Nixon.

In this atmosphere of widening criticism and denunciation of big science, especially manned space flight, preparation for the launch of Apollo 13 went forward.

The "Cyprus" Effect

The landfall for the third mission was a hilly, upland region 110 miles east of the Apollo 12 site called the Fra Mauro formation. The projected landing target was at latitude 3 degrees 40 minutes and 7 seconds south and longitude 17 degrees 27 minutes and 3 seconds west. Compared with the Sea of Tranquility and the Ocean of Storms, the Fra Mauro formation was rough, undulating country with ridges, troughs, and numerous large craters.

The region was named for its most conspicuous feature, the Fra Mauro Crater. From Orbiter photos scientists concluded that the formation was an apron several thousands of square miles in extent consisting of material hurled out of the Imbrium Basin 3.9 billion years ago. It was hypothesized that the basin was gouged out of the Moon by the impact of a planetoid the size of the Island of Cyprus.[1]

Located about 30 miles north of the Fra Mauro Crater, the landing site would put Moon walkers in position to collect samples of material blasted out of considerable depths of the Moon by the Imbrium impact. This might be the original stuff of which the Moon was formed, the primeval crust, excavated, perhaps, from depths of 50 to 100 miles. "In other words, we're using the Imbrium Basin as a big, natural drill hole," Gordon Swann, of the U.S. Geological Survey, told a prelaunch news conference.

Over the eons, however, the ejected material from the great Imbrium hole had been covered by a layer of regolith and dust too deep for explorers to penetrate without elaborate drilling equipment. This difficulty was overcome by locating the landing site near a crater—Cone Crater—that had been excavated by an impact about 3 billion years after Mare Imbrium was formed. Cone Crater, 1100 feet across and 250 feet deep, was much deeper than the regolith and unquestionably penetrated the ancient Imbrium ejecta. This ejecta had, in turn, been splashed out of the crater by impact and much of it—great, heavy blocks—had fallen on the crater rim. If the lunar astronauts of Apollo 13 could land near their target, they could walk up to the rim of Cone Crater and chip fragments off some of those big rocks. Those fragments might represent the primeval stuff of which the Moon was made, the original crust. A traverse of a mile or so to the rim to collect the ancient chips and photograph the interior of the crater was the prime object of the mission, which also included a roster of sophisticated scientific experiments.

Cone Crater was on a ridge about 400 feet above the elevation of the prospective landing target. No one thought the climb would be difficult in gravitational field only one-sixth that of Earth. "They shouldn't really have any trouble negotiating it," said Swann.

Heat Flow

In addition to the instruments that Conrad and Bean had set up in the Ocean of Storms, Apollo 13 explorers were assigned to install a heat flow experiment that would measure the amount of heat rising from the depths of the Moon. The results would indicate whether the Moon actually had a molten core. They would provide new evidence of the Moon's internal structure.

Internal heating of a planet is accounted for in two ways: by radioactive decay of uranium, potassium, and thorium, and by the residue of the original energy of accretion. How much heat was left in the Moon? This determination was probably the most ambitious experiment in Apollo. It was certainly the most difficult for the astronauts to implant.

The astronauts were required to bore two 3-meter (10 feet) holes 2.5 centimeters (1 inch) in diameter and 10 meters (32½ feet) apart. They would use a thumper drill, powered by a 0.5-horsepower motor, which operated like a pneumatic drill used to tear up paving. Probes with precision thermometers and thermocouples were to be emplaced in each borehole. They would measure temperature differences at a 50-centimeter (19½ inches) interval at the bottom of each hole, the absolute temperature at three points near the top, and the thermal conductivity of the soil. This latter measurement would be made by activating (by radio command from Earth) tiny heaters on each probe and recording the temperature rise from one heat sensor to another. In addition, the explorers would bring back a 10-foot core of soil that could be analyzed for heat conductivity in the laboratory.

Four institutions had joined forces in this experiment: the Lamont-Doherty Geological Observatory of Columbia University, Arthur D. Little, Inc., Yale University, and the Massachusetts Institute of Technology.

The experimenters had reported that microwave measurements by radio telescopes on Earth indicated a lunar surface heat flow ranging from 1 millionth to 5.4 millionths of a watt per square centimeter. The upper limit (5.4×10^{-6} watt) nearly equals the heat flow from the Earth itself. Evidence for high temperatures at shallow depths in the Moon comes from surface volcanism, which is inferred from the flooding of the mare basins by lava, from chain craters, which seem to be of volcanic origin, and from rilles, domes, and other symptoms of volcanism. The experimenters calculated that if the abundance of uranium in the Moon is similar to that of the Earth (0.033 parts per million), the interior of the Moon below 250 millimeters would be molten. The heat flow experiment promised to go a long way toward resolving the hot Moon–cold Moon argument.

According to Marcus G. Langseth of the Lamont-Doherty Observatory, the experiment was designed to operate for a year. If the results radioed back to Earth showed an increase in temperature with depth, the experimenters would be able to calculate the temperature at the center of the Moon, he said. Such calculations made on Earth indicated a temperature of 7200°F (4000°C) at the core and 2000°F (1000°C) at a depth of 60 miles—the region where terrestrial lava is formed.

Rubble 20 Miles Deep

The heat flow experiment could be used on any planet on which men could land. Like the seismometers of Apollo 11 and 12, it was a practical tool to look inside a planet and examine its structure. Apollo 13 carried a third seismometer package, and it would greatly enhance the knowledge gained from the previous two. The Apollo 11 instrument failed after 21 days, it will be recalled, but the Apollo 12 seismometers were still going strong. Gary Latham told the Apollo 13 prelaunch news conference that "we now have months of data—three months of it analyzed so far." Of the 90-day analysis, he said, "we've detected 150 'events' we believe to be of natural origin [that is, not produced by the instrument itself]."

The signals—vibrations analogous to sound waves traveling through the lunar rocks and soil—continued to build up gradually to a maximum and then to decay slowly, lasting a long time, compared with signals on the Earth. What did it mean? "The evidence seems to indicate a broken up, rubble structure in the outer 10 or 20 miles," said Latham. "We are not able to say how this structure came into being."

He said it was guessed that there may have been some homogenous material, such as a lava flow, that solidified and then was battered and fractured by meteorite impact over billions of years. Or, he said, "we may be looking at a structure which is the result of the manner in which the Moon was formed . . . by accretion of blocks of material that came together, coalesced to form the lunar body . . . and . . . never did consolidate."

Inasmuch as many seismic signals seemed to be the product of meteoroid impacts (they were all similar to the signal produced by crashing the Apollo 12 LM on the surface), it did not seem that the Moon was seismically active. That is, there were comparatively few Moonquakes.

"We're not moving great blocks of the crust in the way we think we are doing on Earth," he said. "Whether or not there may be a very, very deep molten core, for example, we can't say. It's still possible."

Apollo 13 held out considerable promise for seismic investigators. Not only was it carrying another package of seismometers to another part of the Moon, but Mission Controllers planned to crash the 30,000-pound third stage of the Saturn 5 launch vehicle on the Moon in addition to crashing the 3,000-pound LM ascent stage. The third stage, or S4B, injected the Apollo lunar module into the translunar flight path from Earth orbit. On Apollo 11 and 12, it simply sailed on past the Moon and continued in orbit around the Earth–Moon system until eventually it reentered Earth's atmosphere and disintegrated. Parts might fall anywhere, hopefully in the Earth's oceans. It was far safer and more useful to crash the big S4B on the Moon where its impact should produce seismic waves that would illuminate the Moon's structure to depths of 50, possibly 100, miles.

Odyssey and Aquarius

The Apollo 13 crew dipped briefly into mythology, astrology, and current musical comedy for the call signs of its vehicles. The Apollo spacecraft was named Odyssey and the lunar module, Aquarius. As it turned out, Odyssey was prophetic, for the cruise of Apollo 13 around the Moon was an Odyssean adventure indeed.

The crew commander was Navy Captain James A. Lovell, Jr., 42, a 1952 graduate of the U.S. Naval Academy. A calm, deliberate, and quiet spoken test pilot, Jim Lovell came into the space program with the second group of astronauts in 1962. He and his wife, Marilyn, had four children, two boys and two girls. The oldest, Barbara, was 16 and the youngest, Jeffrey, was 4.

At the time of Apollo 13, Lovell had logged more time in space than any other man—572 hours and 10 minutes. He had flown the record 14-day mission of Gemini 7 with Frank Borman in December 1965, the 4-day Gemini 12 mission with Edwin Aldrin, and the 6-day mission of Apollo 8 to lunar orbit with Borman and William A. Anders.

Lovell's partner-to-be at Fra Mauro, the lunar module pilot, was a civilian test pilot, Fred Haise, Jr., 36, a slender, drawling Mississippian from Biloxi. Haise had a Bachelor of Science degree in engineering from the University of Oklahoma. He was married to a hometown girl, Mary Grant. They had three children, the oldest, a daughter, Mary, 14, and two younger boys, Fred and Stephen.

Before becoming an astronaut in 1966, Haise had been a research pilot at NASA's Lewis Research Center at Cleveland and at the Flight Research Center at Edwards, California. This was his first space mission.

Assigned to remain in lunar orbit aboard Odyssey, the command module pilot was Navy Lieutenant Commander Thomas K. Mattingly, 34, born in Chicago and reared in Miami. He was a 1958 graduate of Auburn University in aeronautical engineering and had been given advanced training at the Air Force Aerospace Research Pilots School.

Five days before the April 11 launch, it was announced that a member of the Apollo backup crew, Air Force Major Charles M. Duke, Jr., had come down with rubella (German measles). Lovell, Haise, and Mattingly were given blood tests to determine if they had immunity, for it would not do to have the disease develop on the flight. Lovell and Haise exhibited immunity, but Mattingly did not. Because he had been exposed to rubella and because it was likely he would be stricken with it on the flight, he was replaced by his back-up, John L. Swigert, Jr., who had immunity. Swigert, 34, was a test pilot with a degree in mechanical engineering from the University of Colorado and a Master of Science degree in aerospace science from Rensselaer Polytechnic Institute, Troy, New York. In his spare time, he acquired a degree of Master of Business Administration from the University of Hartford, but his real love was test piloting. He worked as an engineering test pilot for North

American Aviation (later North American Rockwell) and for the Pratt & Whitney Division of United Aircraft. Aside from his competence as an astronaut, an immunity to rubella, of which he was unaware, got him a couch on Apollo 13.

The Odyssey of the Odyssey

Lovell and Haise were to spend 33 hours on the Moon. They would make two EVAs (extra-vehicular activity periods), the first to emplace the ALSEP* instruments and collect local rock and dirt samples and the second to make a traverse of about 5000 feet to Cone Crater to gather Imbrium ejecta material from its rim. A month before the launch, they had practiced this traverse in the Verde Valley in Prescott National Forest, Arizona, near Cottonwood. There, U.S. Geological Survey geologists had made simulated Moon craters up to 81 feet in diameter by setting off 366 explosive charges.

Thoroughly equipped, rehearsed, and eagerly ready, the crew of Apollo 13 was boosted off Pad 39 A of the Kennedy Space Center on time at 2:13 P.M., Eastern Standard Time, Saturday, April 11, 1970. The launch appeared smooth enough to observers, always thrilled by the thunderous, bone-shaking roar of the mighty Saturn 5, the most powerful rocket in the world. However, all was not quite well. A sequence of minor flaws appeared during powered flight up through atmosphere. They were precursors of, although not connected with, the disaster to come two days later as the crew approached the Moon.

During the firing of the Saturn 5's second stage, the center engine in the cluster of five liquid hydrogen–liquid oxygen engines cut off 132 seconds early as a result of unusually large oscillations in thrust chamber pressure. The guidance computers immediately reacted to keep the remaining four engines burning 34 seconds longer than programmed so that the vehicle would reach planned acceleration at second stage cutoff. Even so, the velocity of the Apollo stack was 223 feet per second lower than planned. Again, the guidance system attempted to compensate by causing the third stage, the S4B, to burn its single hydrogen–oxygen engine 9 seconds longer than programmed. This was necessary to complete the boost of the spaceship into a parking orbit of 100.2 by 98 nautical miles.

When the Apollo lunar module stack, still attached to the S4B, achieved orbit, mission controllers at the Manned Spacecraft Center, Houston, checked the S4B's remaining fuel. Was there enough after this

---

* An acronym for Apollo Lunar Surface Scientific Experiment Package.

extra-long first burn to insert the stack in the translunar trajectory with a second burn? There was plenty, they noted with relief. A great deal of redundancy had been built into the Saturn rocket system, by far the most reliable vehicle in the United States inventory and probably in the world. It never failed during the whole history of Apollo. There were numerous minor malfunctions, but the three-stage vehicle was so powerful and so well engineered that none of its flight problems interfered with its overall performance.

Two hours and 35 minutes after the launch, the crew fired the S4B a second time for translunar injection (TLI). Once on the way out of Earth orbit and headed Moonward, Lovell performed the transposition maneuver. He eased the 100,000-pound Apollo command and service module away from the S4B, turned the stack around, and docked nose first with the lunar module still encapsulated in the forward section of the massive S4B. When the LM was secured to the Apollo command module, the crew activated springs that pushed the LM–Apollo stack away from the S4B. As the Apollo LM moved slowly away from the big rocket, controllers at Houston fired the S4B attitude control thrusters to widen the distance. Then they activated the S4B instrument unit that would steer the 30,000-pound rocket to its lunar crash site. All was well as the rocket and the spaceship went their separate ways. During the evening of April 12, the crew made a course correction that would take them around the Moon at a nearest approach of 62 nautical miles. The new trajectory did not give them a free trip back to Earth, which had been provided as a safety factor in the earlier Apollo lunar flights. If the mission had to be aborted, the crew would have to burn the engine as they rounded the Moon in order to put the ship on a course that would intersect one of Earth's oceans—otherwise it would move in an unstable orbit around the Earth–Moon system.

During the late afternoon and early evening of April 13, Lovell and Haise went forward through a connecting tunnel into the lunar module to check it out. The checkout was performed about an hour ahead of schedule, which was lucky in the light of later events, and all was found to be in order.

Returning to the command module, the astronauts prepared to televise their activities to Houston, but they had trouble getting the high gain antenna into position for signal transmission. It was 8:10 P.M., Central Standard Time, at Houston in the fifty-fifth hour of the flight. Apollo 13 was 174,598 nautical miles from Earth. Jack Lousma, the capsule communicator at Houston, suggested they alter the ship's attitude (position in space relative to the Earth) to obtain better antenna pointing. When this was done, the transmission "locked on." Fred

Haise then gave a demonstration of how he maneuvered from the command module, Odyssey, into the lunar module, Aquarius. It was a bit strange up here, he said, even though he had rehearsed it in a water tank during training. "I find myself now standing, with my head on the floor, when I get down into the LM." What was up in Odyssey was down in Aquarius. In zero gravity it made no difference, but it could be confusing. Haise continued: "The LM, as you can see, looks pretty clean. I found a couple of loose washers about and a little plastic cap off the sequence camera had come loose. . . ."

Lovell appeared in the television picture and Haise explained that if the commander seemed to be standing on a can, it actually was the housing for the LM ascent engine, a very important piece of machinery, inasmuch as it was the only way they could get off the Moon once they had landed.

Haise then demonstrated a "fish scale" weighing device used to determine how much water was left in a bag container. The astronaut attempted to weigh himself on the scale and reported: "It says I weigh actually less than zero right now. Guess its calibration isn't too good." "That will be the day," remarked Lousma at Houston. "I think even you'd weigh zero here, Jack," Haise riposted. "Touché," said Lousma, lightly.

We've Got a Problem Here

After rigging the hammock in the LM to show the folks back home how he would sleep on the Moon, Haise concluded the television show and Mission Control acknowledged that it was "great." Later, Swigert revealed that Haise had spent an hour shaving for the show.

Mission Control announced that ground elapsed time was 55 hours, 47 minutes. Apollo 13 was 177,861 nautical miles from Earth, moving right along at 3,263 feet per second. It was 8:57 P.M.

At 9 P.M., Mission Control asked the crew to roll the vessel to the right about 60 degrees and attempt to photograph a comet named Bennett, which was supposed to be visible then. Lousma was adding further instructions. He asked the crew to "stir up" the liquid oxygen and liquid hydrogen in the service module tanks in order to ensure proper feed to the fuel cell batteries, in which oxygen and hydrogen were mixed to produce electricity and, as a by-product, water.

Suddenly, Haise, who had been talking to Lousma, asked Houston to stand by a moment and then he said: "Hey, we've got a problem here."

LOUSMA: This is Houston. Say again, please.

HAISE: Houston, we've had a problem. We've had a main B bus interval.

There had been a sudden interruption in the ship's electric power.

LOUSMA: Roger. Main B interval. Okay, stand by 13. We're looking at it.

HAISE: Okay, right now, Houston. The voltage is looking good. And we had a pretty large bang associated with the caution and warning (light) there. And if I recall, main B was the one that had an amp spike on it once before.

That is, the amperage on the bus that distributed power to the ship had suddenly surged up and dropped back down.

The crew then reported that main bus A was showing abnormally low voltage. A moment later, it was discovered that liquid oxygen tank No. 2 supplying the fuel cells power system was reading zero. "We are venting something out into space," Lovell reported. "It's a gas of some sort."

It was oxygen. The number 2 tank had ruptured. Two of the Apollo's three fuel cell batteries were dead, leaving only one on line, and this one was showing signs of failing.

Lousma radioed instructions to power down the space ship at once until the trouble could be analyzed. Meanwhile, the vessel began to pitch and roll from the escaping gas.

The Apollo 13 spaceship had suffered a catastrophic accident—the blowout of a liquid oxygen tank in the cylindrical service module behind the command module. It had lost most of its electric power. When this was fully recognized about 10:20 P.M., CST, on the evening of the 13th, Apollo 13 was 180,521 nautical miles from Earth, more than halfway to the Moon. It was hopelessly crippled.

The entire NASA organization was alerted to the disaster. The Public Affairs commentator announced: "Here at Mission Control we are now looking toward an alternate mission swinging around the Moon and using the lunar module power systems because of the situation that has developed this evening."

It was the understatement of Project Apollo. It meant that Lovell, Haise, and Swigert would have to use the electric power and life support system in the lunar module—their lifeboat—to return to Earth alive. There was still some power from oxygen tank No. 1 supplying one fuel cell on Apollo, but the tank was leaking, apparently damaged by the blowout of tank No. 2, and the power was fading. Only the command module storage batteries were left in Apollo—and these would be needed for reentry when the disabled vessel returned to Earth. The

lunar module was the crew's only salvation. Had this accident happened on the flight of Apollo 8, when Lovell went to lunar orbit with
Borman and Anders, the crew would have been doomed, for there was
no lunar module on Apollo 8.

Exploration of an Accident

The main electric power system for the Apollo spacecraft was contained in Sector or Bay No. 4 of the service module, the cylindrical
portion of the vehicle aft of the conical command module where the
crew rides. Power at 28 volts direct current was supplied by three fuel-
cell power plants. In these, oxygen and hydrogen react to produce
electricity, with water as a by-product.

Oxygen was stored in a semiliquid, semigaseous (supercritical) state
in two tanks mounted side by side on a shelf in Bay 4. Hydrogen, in the
same state, was stored in two tanks below the shelf.

Each of the oxygen tanks was a 26-inch sphere of inconel metal
(nickel–steel alloy), double walled with insulation between the shells
to prevent outside heat from leaking into the supercold fluid within.
Oxygen becomes liquid at 297°F below zero. So well insulated were
the tanks, North American Aviation informed the press, that an ice
cube placed inside would take 8½ years to melt.[2]

In the zero gravity of free fall in space, fluids do not flow unless
pushed by something. The method of forcing the oxygen to flow into the
fuel cells was to boil it to build up gas pressure in the tank. For this
purpose each tank contained heater coils, controlled by a thermostatic
switch, and electric fans to stir up the fluid–gaseous mix and promote
an even flow. The thermostatic switch was designed to open and shut off
the heating elements when the temperature rose above 80°F.

Each of the oxygen tanks thus had the same fire and explosive potential that had resulted in the tragic Apollo 1 fire of January 27, 1967.
The recipe for disaster was the same—a mixture of high-pressure oxygen, electric wiring that could, in the event of a short circuit, emit a hot
arc, and insulation that would burn rapidly in pure oxygen at high
pressure. Each tank held 326 pounds of oxygen under pressure of 865
to 935 pounds per square inch. Although each had a burst pressure of
2200 pounds per square inch, and a relief valve that would dump
oxygen overboard when the pressure reached 1000 pounds, these safety
devices could not prevent an explosion from a rapid build-up of pressure if a fire was ignited in a tank. And that is what happened to
Oxygen Tank No. 2 in the Apollo 13 service module 2 days out from
Earth. It was the fatal fire Apollo 1 repeated on a smaller scale—a

short circuit, a hot spark, and the ignition of plastic insulation and metal in a high-pressure oxygen atmosphere. This time, the tank exploded, blowing out the Bay 4 panel on the hull of the service module. That was the "bang" the crew heard.

Fortunately, the hydrogen tanks below the oxygen shelf remained intact. If they had ruptured, the entire vehicle might have been destroyed and Apollo 13 would have become a $400-million coffin, fated to circle the Earth–Moon system until its orbit decayed and it fell into Earth's atmosphere to be incinerated.

As it was, the explosion of Tank No. 2 damaged Tank No. 1, which sprang a leak. Two of the fuel cell power generators supplied by Tank No. 2 died immediately and the one remaining generator lost power as its oxygen supply dwindled from Tank No. 1.

The Strange History of Tank No. 2

The root of this accident can be traced back to 1966, when Tank No. 2 was manufactured by the Beech Aircraft Corp., under a subcontract from North American Aviation, the prime contractor for Apollo.

According to NASA, acceptance testing showed that heat was leaking into the tank at a higher rate than specifications permitted.[3] The rate of heat leakage was reduced after some reworking, but it still exceeded specifications, the agency reported, when the tank was finally accepted with a formal waiver of this discrepancy. Several other discrepancies that were regarded as minor by space agency inspectors were also accepted, according to NASA. These included oversized holes in the tank dome electrical plug support and an oversized rivet hole in the heater assembly just above the lower fan. None of these items was regarded as serious in 1966 and none had anything to do with the explosion 4 years later. But they were symptomatic of a tendency toward oversight in which a more serious discrepancy could occur, undetected.

After it was shipped to North American, Tank No. 2 was first installed in service module 106 for the flight of Apollo 10. On that mission, Air Force Colonel Thomas Stafford and Navy Commander Eugene A. Cernan tested the lunar module, Snoopy, in lunar orbit, descending to an altitude of 50,688 feet from the lunar surface. It was the final test before the landing of Apollo 11.

However, Tank No. 2 was diverted from this flight because of a decision by NASA to modify vacuum pumps on the tank dome. The modification required the removal of the cryogenic (super-cold) oxygen tanks and the shelf on which they were mounted in the Service Module. As Tank No. 2 was being removed, it was inadvertently dropped about two inches. Subsequent tests indicated that the jarring apparently had not caused any damage to the interior assemblies. Following vacuum pump modifications, the tank was installed in service module 109 for the flight of Apollo 13.

Weeks before each Apollo launch at the Kennedy Space Center, a countdown demonstration test series is carried out to detect any problems in the Saturn–Apollo lunar module array before the final countdown begins. During the countdown demonstration tests in March 1970, ground crews reported a problem in Tank No. 2. Although it could be filled normally, it could not be emptied in the normal manner —by pumping gaseous oxygen into the vent line to push the liquid oxygen out through the fill line. This procedure worked perfectly for Tank No. 1, but not for Tank No. 2.

An analysis by the ground crews indicated the possibility of a loose fitting that allowed the gaseous oxygen being pumped in the vent line to escape through the fill line without pushing out much liquid oxygen in the tank. Later, the possibility that the fitting had been loosened when the tank was dropped at North American months before was considered as the cause of the problem.[4]

The ground crew resorted to a simple expedient to empty the tank— by turning on the heaters and fans to boil the oxygen out. On March 27 and 28 the heaters and fans were turned on by applying 65 volts of direct current from the ground power supply for periods of 6 and 8 hours.

Unknown to the ground crews, consisting of both NASA and contractor personnel, this set the stage for the accident. The 65 volts were much too high for the thermostatic switches that controlled the heaters. These switches had been designed to operate on 28 volts of direct current from the spacecraft fuel cell generators. Although the switches would carry 65 volts when closed, they would fail in the closed position if they started to open to interrupt the load, according to the findings of the NASA Review Board. At one point, the Board concluded, the switches did start to open and then were welded shut during the long period when the heaters were energized with 65 volts to boil the oxygen out of the tank.

"From that time on, including pad occupancy, Oxygen Tank No. 2 was in a hazardous condition when filled with oxygen and electrically powered," the NASA investigation reported.[5]

The failure of the thermostatically controlled switches allowed temperatures to reach 1000°F in the heater tube assembly (instead of shutting the heaters off at 80°F) during the countdown demonstration test. The intense heat burned off the Teflon insulation on the fan motor wiring, the investigation concluded. The wires were bare. They could have short circuited at any time after that—even while Apollo 13 stood on the launch pad atop the Saturn 5 launch vehicle fueled with 215,330 gallons of kerosene in the first stage, 272,340 gallons of liquid hydrogen in the second stage, and 64,145 gallons of liquid hydrogen in the third stage. For three weeks before its launch, Apollo 13 was a potential bomb, with the energy of a World War II "block buster."

A Persistence of Oversight

Had Tank No. 2 exploded while Apollo 13 was on the pad, "the entire launch vehicle might have been destroyed," according to Edgar N. Cortright, director of NASA's Langley Research Center, who headed the Apollo 13 Review Board.[6] A second Apollo disaster on the launch pad would have terminated the program.

In the 1967 Apollo 1 fire, Teflon insulation and other materials caught fire from a hot electric arc of a short circuit in the command module where the crew was running a series of prelaunch tests on the spacecraft. Apollo 1 stood atop its Saturn 1B launch vehicle, fueled for a maiden test flight in Earth orbit. It was late afternoon of January 27 when the crew shouted "fire" over the communications system, which then went dead. Unable to extricate themselves from the smoke-filled cabin, the three astronauts were suffocated. They were Air Force Lieutenant Colonels Virgil I. (Gus) Grissom and Edward H. White II, and Navy Lieutenant Commander Roger B. Chaffee. The Saturn 1B launch vehicle was not affected because the fire was confined to the command module.

The conditions in which the Apollo 1 and 13 accidents occurred betray persistent oversight. In both cases, the possibility of a short circuit and its effects in a pure oxygen environment were not even considered. NASA and its contractors had taken safety precautions to prevent an Apollo 1 accident from recurring in the cabin of the command module, but not in the service module.

The pure oxygen atmosphere used in Mercury, Gemini, and the Apollo 1 cabins was modified to a less flammable cabin atmosphere of 60 percent oxygen and 40 percent nitrogen in all subsequent Apollos to minimize fire hazard during launch. Then, during the early hours of the mission, a waste management dump valve was left open to allow this mixture to leak out slowly and be replaced with pure oxygen. The result was a gradual oxygen enrichment in the cabin atmosphere during the mission, although a very small percentage of nitrogen remained throughout most of the flight.

Unaware of the short-circuit hazard in Tank No. 2, NASA and contractor engineers had considered replacing the tank because of the emptying problem, but rejected the idea after further tests March 30 showed the problem, believed to be caused by a loose fill tube sleeve, would not affect tank operation in flight. It is probable that pressure to get on with the flight and avoid a month's delay, which changing the tank would have required, had something to do with this decision. In the light of known factors, the decision probably was right. In the light of hindsight, it was wrong.

Why did the ground crew apply 65-volt power to a circuit controlled by a 28-volt safety switch? The original 1962 specifications from North American to Beech Aircraft Corporation for the tank and heater assembly specified the use of 28-volt direct current, as used in the spacecraft. In 1965, North American issued a revised specification. It stated that the heaters should use a 65-volt direct current power supply for tank pressurization. The Review Board explained that this was the power supply used at Kennedy Space Center to reduce pressurization time. The Board reported: "Beech ordered switches for the Block II tanks but did not change the switch specifications to be compatible with 65 volts direct current." The discrepancy was not detected by NASA, North American, nor Beech in their documentation reviews. The switches were not given qualification and acceptance tests as should have been done, the Review Board said. Thus, an incompatibility between the load the switches could carry and the Kennedy Space Center ground power supply existed. "It was a serious oversight in which all parties shared," the Review Board concluded.

Presumably, the Apollo 13 accident could have happened on any of the lunar flights, including Apollo 8, which flew without the LM, if those vehicles had been subjected to the detanking procedures that the ground crew used on balky Tank No. 2 in Apollo 13. But the need to apply 65-volt ground power to a tank heater long enough to cause a safety switch to fail did not arise until the Apollo 13 countdown demonstration test. And it might not have arisen if the fill tube fitting had not been loose—and this fitting might not have been loose if the tank had not been dropped at North American.

Nevertheless, the fact that the safety switches had been welded shut and hence were not operating could have been detected at the Kennedy Space Center if someone had been watching heater current readings on Tank No. 2, the Review Board said. The indicators would have shown that the heaters had exceeded the safety switch temperature limit.

The blueprint for the accident was finally drawn by the Review Board. Because of a bump one day in the fall of 1969, a fitting might have been loosened. Because of that, a tank could not be emptied properly. Because of that, a ground crew applied the wrong voltage to the tank heaters. Because of an inadequate switch, overheating occurred, burning insulation off electrical wiring. Because of that, the wires eventually short circuited and a $400-million mission was aborted.

Because of a nail, the battle was lost.

Salvation

During the first 46 hours of the mission, as the Review Board reconstructed it, all went well. At 46 hours and 40 seconds ground elapse time, the crew turned on the fans in Tank No. 2 as a routine procedure. Something happened. Within 3 seconds the oxygen quantity reading jumped from 80 percent filled to an off-the-scale high reading of more than 100 percent.

That was when a short circuit occurred, the Review Board concluded. But, other than the off-scale reading in the cabin, there was no immediate indication of it.

An hour later, the fans were turned on again. And 3 hours after that, they were turned on a third time. The quantity gauge continued to read off-scale high. Was the gauge malfunctioning? No one could be sure.

At 55 hours, 52 minutes, and 30 seconds into the flight the master alarm on the spacecraft's caution and warning system alerted the crew to low pressure in Hydrogen Tank No. 1. This was noted also at Houston. Jack Lousma called up the ship and told Lovell to turn on fans and heaters in the cryogenic system.

About 90 seconds after this was done, spacecraft telemetry signals to Houston blacked out for 1.8 seconds. During the blackout, the spacecraft caution and warning system alerted the crew of low voltage on main bus B from which power is distributed to the spacecraft. At that moment, the crew heard a loud bang. It was the explosion of Tank No. 2, which also blew out the panel covering Bay No. 4 and damaged Tank No. 1.

As NASA summarized the situation later: "The resultant loss of oxygen from both tanks made the three fuel cells inoperative, leaving the command module with storage battery power used during reentry and only a small supply of oxygen in the surge tank used to repressurize the cabin after it was vented. The Lunar Module therefore became the only source of sufficient electrical power and oxygen to permit the safe return of the crew to Earth."

In its summary review of the accident, the Review Board also stated on June 15, 1970: "It is now known that the tank contained two protective thermostatic switches on the heater assembly that were inadequate and would subsequently fail during ground test operations at the Kennedy Space Center."

In addition, the Board said, it was probable that the tank contained a loosely fitting fill tube assembly, displaced during handling, possibly from the 2-inch drop. While the displaced fill tube in itself was "not particularly serious," the Review Board continued, it led to "improvised detanking procedures" at the Kennedy Space Center setting up the conditions for the explosion.

There were two reasons why NASA and contractor people did not realize the hazard of using the heaters for long periods to empty the tank. Some were not aware of the extended use of the heaters. Others who knew about it assumed that the safety switches would prevent overheating. No one at that time was aware that the switches were not designed to function under a 65-volt load.

Return to Earth

As the full extent of the disaster became apparent aboard the spaceship and at Houston, the crew began moving into Aquarius and closing down the command module, Odyssey. Navigators at Mission Control computed a new course that would swing the crippled vessel around the Moon and bring it back safely to Earth. At 2:43 A.M., CST, April 14, the first course correction was made. The LM's powerful descent engine was fired for 30.7 seconds through the module's Abort Guidance System. The firing boosted velocity 38 feet per second, which put the spacecraft on a free return trajectory that would intersect the Indian Ocean near the British Island of Mauritius, off the East Coast of Africa in 90 hours and 30 minutes. On that path, splashdown would come at 9:13 P.M., Houston time, April 17.

On its new course, Odyssey–Aquarius would pass the Moon at a closest approach of 136 miles, instead of 68 miles on the earlier course. Jack Lousma, capsule communicator at Mission Control, relayed this computer information to the crew, saying:

"And, Aquarius, for your information, we now have a 136-mile perigee."

"Wait a minute, Jack!" Lovell shot back. "Did you say pericynthion or perigee?"

"I mean pericynthion," replied Lousma.

"That's better."

Perigee refers to a point in orbit closest to Earth. The analagous term for a Moon orbit is pericynthion or perilune. A 136-mile perigee would mean that they would miss the Earth.

The little slip relieved some of the tension in space and on the ground. The situation was bad, but the prospects that Lovell, Haise, and Swigert would survive it were good. Not excellent at that time—just good. The prognosis at that "point in time" as the engineers (and later White House staffers) liked to say was "fair." Outwardly, though, everyone appeared calm and confident.

As Odyssey–Aquarius neared the Moon, Mission Control devised a second course change. Two hours after reaching pericynthion, the crew was advised, Aquarius' descent propulsion engine was to be burned for 263.4 seconds. This would increase velocity by 861 feet per second. It would cut 9 hours and 6 minutes from the return time and put the splashdown in the mid-Pacific Ocean at 7 minutes past noon, Houston time, April 17. This splashdown point would make it simpler to recover Odyssey, since the recovery ships already were deployed in the Pacific.

This final course correction burn was scheduled at 79 hours, 28 minutes ground elapsed time, or 8:41 P.M., April 14. Apollo 13 rounded the Moon in its 77th hour of flight and headed back toward Earth at 7,064 feet per second.

"That view back there is fantastic," said Fred Haise. "You can see where we're zooming off."

"Yes, we're no longer 139 miles," said Lovell. "We're leaving."

Later, the capsule communicator, Vance D. Brand, advised: "By the way, Aquarius, we see the results now from 12's seismometer. Looks like your booster just hit the Moon and it's rocking it a little bit. Over."

"Well, at least something worked on this flight," said Lovell. "I'm sure glad we didn't have an LM impact, too."

At Mission Control, very important spectators arrived to observe the moment of truth—the firing of the LM descent engine. The viewing room was filled. There were NASA Administrator Paine and his deputy, George Low; Frank Borman, who had flown Gemini 7 and Apollo 8 with Jim Lovell; Ken Mattingly, who had been bumped from the flight of unlucky 13; and Alan B. Shepard, the first American to fly in space and now the commander of the next Moon flight, Apollo 14, along with his crew mates, Stuart Roosa and Edgar Mitchell.

There was a long silence in the big room. Then Lovell's voice came over the communications system announcing engine shutdown.

"I say that was a good burn," the capsule communicator said.

"Roger," said Lovell in a matter-of-fact way. "And now we want to power down as soon as possible."

Apollo 13 was coming home free and all the crew had to do was survive another 63 hours and 22 minutes until they were required to bring Odyssey down into the Pacific Ocean.

Actually, they did more than that. With instructions from Houston, they rigged an air purifier to remove excess carbon dioxide. They made observations and talked at length with Earth, crowded as they were in the tiny cabin of Aquarius. Five hours out from the home planet, they burned Aquarius' reaction control thrusters for 22.4 seconds to adjust their reentry angle. They then separated the command module from the damaged service module, using Aquarius' thrusters to pull both lunar module and command module away from the big, dead cylinder. For the first time, they saw what the explosion had done as the service module drifted away.

"There's one whole side missing," Lovell said. "The whole panel is blown out, almost from the base to the engine. It's really a mess." Strips of insulation were hanging out of the wrecked bay.

The crewmen then crawled back into the command module and sealed off the tunnel leading to Aquarius. They were finished with the lifeboat which, having no heat shield, could not alight on a planet with an atmosphere. Powering up Odyssey with its storage batteries, they prepared for a routine reentry, first undocking from Aquarius. The little lifeboat retreated rapidly from Odyssey, propelled away by the whoosh of its cabin oxygen escaping into space through the open hatch.

Odyssey plowed into the atmosphere, toward the blue ocean below. It splashed down at 7 minutes, 41 seconds after noon, Central Standard Time, April 17, in sight of the U.S.S. *Iwo Jima.* In less than an hour, the crew had been extricated and flown to the deck of the carrier by one of its helicopters.

The worst journey in Project Apollo was over and all hands were ashore.

There was only thing left to say and it was said by Robert R. Gilruth, one of the pioneers of the aerospace age, then director of the Manned Spacecraft Center, Houston: "I think it has been made quite clear, more than by any of the Apollo flights up to this point, that flying to the Moon is not just a bus ride."

Aftermath

The safe return of the Apollo 13 crew was hailed as a great achievement, a true victory of man over adversity. But after a national wave of relief subsided, the failure of the mission added fuel to fires of criticism that repeated voyages to the Moon were wasteful, unnecessary, and irrelevant.

Aquarius vanished and probably burned up as it fell into the atmosphere. However, the 8.36 pounds of plutonium-238, a highly poisonous radioactive isotope, which was carried aboard the lunar module to provide power for the ALSEP experiments in a SNAP 27 thermoelectric generator, probably did not burn up. It is probably lying on the bed of the Pacific Ocean, sealed in its graphite cask which, the space agency said, was designed to withstand reentry heating "in the unlikely event of an aborted mission."[7]

Wherever that plutonium is, and it is the largest supply of lost plutonium on this planet, it will continue to generate heat and alpha-particle radiation for centuries, degrading and eventually breaking down its container. When its containment has been breached, it will become a source of radioactive pollution—possibly far from human habitation in the Pacific, but exactly where, no one knows.

The political impact of the Apollo 13 debacle was not felt immediately, but NASA officials sensed it developing and hastened to reassure their friends in Congress that the accident was not so catastrophic as it appeared and did not cloud future flights.

Addressing the Senate Committee on Aeronautical and Space Sciences on April 24, 1970, Paine referred to the oxygen tank as a mere "oxygen bottle" or "thermos flask." The problem it posed could be readily solved, he said.[8] But the tank was more than a thermos flask. It was a complex, cryogenic, oxygen storage unit. What Paine did not tell the Senate space committee, because he did not know it then, was that the "fix" for this "oxygen bottle" in Apollo 14 and subsequent missions was going to cost $15 million and delay the entire program four months.

At that time, Paine was saying that necessary alteration should be made promptly, and it should be possible to get on with lunar exploration on approximately the same schedule as before.

Paine, despite his optimism, realized that NASA could not afford another Apollo 13. He told the committee: "Finally, let me state my view as to how the Apollo 13 accident should affect the future of the space program. I see no reason why this setback should be, or should be made to be, the occasion for a major change in the course of the nation's space program. We have clearly demonstrated the basic soundness of the Apollo system. . . ."

The real threat to continued lunar exploration, however, was the progressive public disenchantment with the space program, not any particular failure. The war in Southeast Asia, which had been eating into the body politic in America like a cancer, was contributing to the decay of public confidence, especially among the huge World War II and immediate postwar generation entering adulthood, in the Nixon administration, and in Congress. The hostile response to the war policy bred a similar response to other quasi-military programs—and the space effort was regarded by all of its critics as such a program.

By the middle of 1970, the space program, which had carried out the most brilliant technological feat in history, was regarded by the rising generation of young skeptics with the cynicism that their parents had viewed the leaf-raking projects of the old WPA in the 1930s and 1940s. The term "Moondoggle" would not go away. In vain did NASA propaganda proclaim the value of space technology as the cutting edge of new invention, of new industrial processes. All this was true but of very little consequence to an electorate obsessed with the problems of war, inflation, the draft, the deserters and evaders, student protests, riots, and race tension.

Between the triumph of Apollo 11 and the debacle of Apollo 13, hardly 9 months had elapsed, but it was enough time for the gestation of a new, negative public attitude. After the rescue of the Apollo 13 crew, public attention turned away from the space program. Lunar exploration went forward under its own momentum. No one on Capitol Hill, it seemed, was pushing it very hard.

Following the report of the Apollo 13 Review Board in June, Paine announced that fixes in the service module would cost $10 to $15 million and that Apollo 14, the first scheduled for launch in October, would be delayed until the end of January 1971.

A story datelined July 7 in *The New York Times* reported that the Senate "narrowly rejected two attempts by liberals to cut the nation's space budget as part of a campaign to re-orient government priorities toward everyday human needs."[9] What the liberals had failed to do was to obtain a reduction in the $3.3 billion NASA was asking for the 1971 fiscal year.

Senator William Proxmire, Democrat of Wisconsin, offered an amendment that would cut the NASA appropriation $122 million. This was defeated, 38 to 35. Proxmire, a strong, critical voice on Capitol Hill, was an ardent foe of large expenditures for science and technology. High on his list of "wasteful" projects, next to Apollo and other NASA ventures, was the U.S. supersonic transport.

Going Proxmire one better, Senator J. William Fulbright, Democrat of Arkansas, put in an amendment to cut $300 million from the space budget. He lost, 37 to 32.

Fulbright was disturbed because the Nixon administration had vetoed appropriations providing a half-billion dollars in federal aid to local governments for sewers and waterworks. Fulbright contended that if the administration wanted to hold the budget down, space should be cut, not sewers, which the people needed more than Moon flights. There was little argument about that. However, fellow senators recognized pork barrel when they saw it, and sewer money for municipalities and counties to finance storm and sanitary sewers was in the old tradition of pork barrel. Even so, no one could refute Fulbright's thesis: "It is a matter of reordering priorities and sewers are more important than more rocks from the Moon."

Unhappily for science, this was the logic of the counter-scientific revolution that threatened to knock the props from Apollo.

NASA was not without a defender. Senator Gordon Allott, Republican of Colorado, reminded his colleagues that the lunar missions had provided "a great thrust forward" in scientific activities in many fields.

There were other instances of the growing political attack on NASA, which spread after Apollo 13. Early in June, during the New York State gubernatorial primary election campaign, Arthur J. Goldberg, a candidate for the Democratic nomination for governor, promised not only to overhaul the state budget but to do what he could to scrap the Apollo program as a means of cutting the national budget. Goldberg, a former Chicago lawyer who had become first Secretary of Labor and then an associate justice of the Supreme Court in the Kennedy administration and the United States representative to the United Nations in the Johnson administration, was one of the most influential liberals in the Democratic party. He was quoted as saying during his campaign that NASA had enough lunar rocks to answer specific questions about the origin of the Moon.[10]

Then, on September 2, 1970, the space agency announced that two more lunar landings had been cut from the Apollo program, reducing the total number of landing attempts from the original ten to seven. Of the seven voyages of Apollo, two—Apollo 11 and 12—had been flown successfully, and one—Apollo 13—had failed to land. That left four more landing missions: Apollo 14, 15, 16, and 17.

The retrenchment stunned the Apollo community of scientists, contractors, and astronauts. Twelve days after the announcement, Administrator Paine left NASA to become vice-president of the General Electric Company. He said that the cutback had not motivated his resignation. The White House was giving way to the sewers-not-space species of argument in attempting to hold its budget line against spending pressure in Congress and, for a while, it looked as though the Apollo program would be dismantled.

Reaction from the scientists involved in Apollo was quick, but in-effective. "We will not get our first order scientific answers from Apollo" as a result of the cuts, said John W. Findlay, chairman of the NASA Lunar and Planetary Missions Board.[11]

Said Thomas Gold, astronomer at Cornell University and an occa-sional critic of manned flight: "It's like buying a Rolls Royce and then not using it because you claim you can't afford the gas."[12] It was now unlikely that answers to the questions of the origin and evolution of the Moon, the Earth, and the Solar System could be found from Apollo with nearly half of its remaining flights cut out, Gold added.

NASA estimated that the cutback would save $20 million per flight in operational costs. Such a saving was "chicken feed" in view of the $25 billion already spent on Apollo, said Harold Urey at the University of California, La Jolla. Thirty-nine scientists sent a joint protest to the space committees of the Senate and the House of Representatives. They said that the program cuts would eliminate one-third of the science planned for the Apollo series.

In a letter published in *The New York Times*, Cal Tech's Wasserburg commented that "it is particularly difficult to understand the justifica-tion for the cancellation of Apollo 19 which was to have performed the most advanced experiment. . . . This decision threatens our existing investment in planetary exploration during a period when we are obtain-ing maximum returns and makes doubtful the scientific justification of the manned space program. The total effect seems to indicate a return to the dark ages of planetary science."[13]

The New York Times reacted in an editorial September 4 entitled "Retreat From the Moon." It observed that: "An incredibly intricate technology and the elaborate organization built to exploit that tech-nology are, in effect, being abandoned . . . now that the easily bored world audience has begun to yawn."

Perhaps the irony of the cutback was expressed most succinctly by Richard D. Lyons, one of the reporters covering the program for *The New York Times*. "When American politicians, goaded by Soviet space successes, first proclaimed the national effort to land men on the Moon, many of the country's most influential scientific leaders scoffed," he wrote. "At least four science advisors to the President raged against the Apollo program, calling it a scientific luxury, a weight lifting contest with the Russians and a plan that wouldn't work. A decade later, the scythe of discontent with Apollo has swung full circle. Now it is the politicians who are attacking the program while the scientists wring their hands over the lack of support."[14]

The contest between money for sewers and space continued in the

Senate. On December 7, the Senate rejected 50 to 26 an amendment to the NASA appropriation bill by Senator Walter F. Mondale, Democrat of Minnesota, to cut $110 million for development of the space shuttle,* the main post-Apollo project in manned space flight. Mondale described the shuttle as "one of the most indefensible items in this budget" particularly in the light of domestic needs.

Defeated also, 52 to 25, was an effort by Fulbright to add $150 million to the pork barrel appropriation for local sewer construction. Again, Fulbright argued that sewers and waterworks were a lot more essential to the American people than manned space flight programs.

A Point of Relevance

The 10-month recess in lunar flights between Apollo 13 and 14 pulled the Moon program down to a low point in public attention. NASA organized a news conference in Washington, November 18, 1970, in the hope of regenerating public interest. Apollo 13 had yielded a scientific result after all, as the crew of Apollo 13 had been told. The booster rocket, S4B, weighing 30,700 pounds, had crashed on the Moon at 7:09 P.M., Houston time, April 14, just 24 nautical miles from the Apollo 12 seismometer. And the Moon rang from that blow for nearly 4 hours. New seismic data were indicating the possibility of a very thick lunar crust.

Although the November conference did not achieve big headlines, the presentation of Frank Press, who spoke on seismic results, deserves to be remembered. It conveyed the conviction of most Apollo scientists that the program had to go forward as an essential part of mankind's intellectual growth and understanding of his environment.

"The basic proposition involved in lunar and planetary exploration," Press said, "is to understand how the planets evolved from the original cloud of dust and gas and clumps of matter to protoplanets, which then accreted to form the final planets as we know them . . .

"And then how each planet separately followed its own evolution over billions of years to the present. No matter how we study the Earth from the Earth, we end up only with a description of what happened,

---

* The shuttle will be a space airplane about the size of the Boeing 707. It is being designed to fly weekly Earth orbit missions by 1980 and is the primary component in a future manned, interplanetary transportation system replacing Apollo. The aerospace craft is to be boosted off the pad with the aid of rockets which are then dropped in the ocean and recovered for re-use. It then flies into orbit under its own power. On the return, it de-orbits under power and then coasts to a landing on a conventional airport runway. Ultimately, it will replace most of the launching rockets in NASA's inventory.

rather than an explanation. It takes the exploration of other planets to provide the deep understanding which is basic to knowing planetary processes . . . not only what is happening, but why."

What was the "relevance" of knowing planetary processes? Press had an answer. "At the present time," he said, "the way we prospect for minerals is to go out and take our chances and look for them. We might use electronic methods, but it's still a 'hunt and pick' method. There is yet no theory which explains why these deposits are where they are."

If such a theory could be formulated, he implied, mining engineers and geologists would know where to look. But beyond that:

"The exploration of the Moon is part of the great intellectual quest of man to explore the cosmos, a drive that he has had since the beginning of time, not only to protect and enrich himself but essentially to satisfy his desire to probe the unknown.

"It would be a pity if the exigencies of the moment deflected us from this very important exploration, perhaps the most important exploration in the history of man."

The year 1970 was a bad one for NASA. At the start, hopes were high, but then disaster struck and the magnificent lunar adventure began to crumble.

At the end of it, the Fra Mauro formation still beckoned.

Notes

1. Ganapathy, R., Laul, J. C., Morgan, J. W., and Anders, E., "Possible Nature of the Body That Produced the Imbrium Basin." *Science*, January 7, 1972.
2. Apollo News Reference, published by North American Aviation, Inc., 1969.
3. Report of the Apollo 13 Review Board, NASA, June 11, 1970.
4. Ibid.
5. Ibid.
6. *The New York Times*, June 16, 1970, report of NASA News Conference, June 15.
7. Apollo 13 Press Kit, NASA, 1970.
8. Statement of Thomas O. Paine before the Senate Committee on Aeronautical and Space Sciences, April 24, 1970.
9. *The New York Times*, July 8, 1970.
10. *The New York Times*, June 5, 1970.
11. *The New York Times*, September 3, 1970.
12. Ibid.
13. *The New York Times*, September 13, 1970.
14. *The New York Times*, September 6, 1970.

8

Fra Mauro

Following modifications to the Apollo service module's oxygen supply system costing $15 million, the space agency prepared to resume lunar exploration with the launch of Apollo 14 on January 31, 1971. The target was still the Fra Mauro formation where program scientists hoped the astronaut explorers would find the rocks that had made up the primeval lunar crust.

Although the fixes in Apollo were tantamount to closing the barn door after the horse departed, the reworking of the oxygen system added more capacity and made it possible to enlarge operations of later missions. It was NASA's plan to extend the stay in lunar orbit and on the Moon in the last three missions, Apollo 15, 16, and 17, and also to increase the number, complexity, and weight of the experiments and exploration equipment.

A third oxygen tank was installed in an empty bay of the service module. An auxiliary battery was added as a backup in case of fuel cell failure. The fans and thermostatic switches were taken out of the oxygen tanks. Wiring inside the three tanks was encased in stainless steel (instead of the previous Teflon insulation that was flammable). The manner of running the heaters was changed to avoid inadvertent overheating and sensors were installed that would report tank temperatures to Houston as well as to the astronauts in the command module. An emergency 5-gallon supply of drinking water was added to the command module.

On the eve of the launch of Apollo 14, the principal investigators were waiting for the other shoe to drop. The maria had been sampled. What were the highlands like? Surveyor 7 had provided an important clue, but samples that could be analyzed in the laboratory were needed.

Apollo 11 and 12 had shown the nature and probable evolution of the maria. Preliminary data indicated episodes of melting over a period of a billion years, but this interval that suggested the length of time volcanism had gone on was disputed. Preliminary dating of some rocks returned from Oceanus Procellarum by Conrad and Bean yielded ages of 2.6 to 2.9 billion years for the formation of the mare. Since the Apollo 11 rocks had been dated at 3.6 and 3.7 billion years, the basalts in Procellarum appeared to be a billion years younger, indicating an entire eon of volcanism.

At the spring 1970 meeting of the American Geophysical Union, however, Wasserburg and his colleagues in the "lunatic asylum" at the California Institute of Technology reported that their analyses of Apollo 12 basalts showed a maria age of 3.4 billion years. This result would shorten the apparent volcanic period to 200 or 300 million years. Later investigation, however, would expand it again, to 550 to 800 million years. The elasticity of dating results was somewhat confusing, but it produced interesting arguments.

Now the time had come to approach the brighter parts of the Moon, the so-called terrae (continents) or highlands. The Fra Mauro formation was at the southern edge of the Imbrium Basin, and it appeared to be a source of very ancient continental rocks, which perhaps had been the original crust.

Rock fragments that differed in chemistry and age from the predominant basalts had been found in both Tranquillitatis and Procellarum. These erratic fragments were older than the basalts and possibly had been thrown out of the highlands.

One of the first reports that the fragments represented a low-density crustal rock of the type that would have formed a surface scum on denser, molten material was submitted to the Apollo 11 Lunar Science Conference by John A. Wood and his associates of the Smithsonian Astrophysical Observatory.[1] In examining thin sections of a bulk sample that Armstrong and Aldrin returned from Tranquility Base, the Smithsonian group found 61 grains of material they called gabbroic anorthosite (or anorthositic gabbro*). The material was low in iron and titanium compared with basalt at Tranquility Base and lighter in color. Its density was only 2.9 grams per cubic centimeter, compared with the whole Moon average of 3.4.

Significantly, its bulk chemical composition matched that of highland surface material reported by the Surveyor 7 alpha back-scattering experiment in 1968.

---

* Gabbro is a general term for a granular igneous rock.

"Very probably our anorthosites are highland fragments tossed into Mare Tranquillitatis by crater impacts," the Smithsonian report surmised. "The simplest model of near surface lunar structure consists of a light anorthositic crust floating on denser Ti [titanium] gabbro. Maria are giant holes in the crust, torn out by major cratering events into which basalt has welled from the substrate."

The assumptions of this model were a long way from being proved, but the existence of a rock type in the highlands that might have formed the primitive crust before it was smashed up by bombardment seemed to be a strong possibility. And Fra Mauro seemed to be just the place where it was likely to be found. Photographs from both Apollo and the earlier Orbiter reconnaissance suggested that the Fra Mauro formation was composed of material heaved out of the Imbrium Basin by the impact that formed it. In this hilly region, therefore, might lie the earliest rocks of the Moon, rocks excavated out of the lunar crust and piled up on the basin's rim.

A Lunar Norumbega

At this point in the exploration of the Moon, parallels can be drawn with the European explorations at the fringe of North America in the late fifteenth and early sixteenth centuries. Although the motives underlying these two periods of exploration are not readily comparable, there are similarities in expectations and in viewpoints.

Starting with Columbus in 1492 and John Cabot in 1497, European mariners had as their principal objective a passage to India and China. They were not searching for the new world they found. Nor were the scientists and astronauts in Project Apollo seeking a new world. They were hunting on the Moon the answers to basic questions of the origin and evolution of the Solar System, which the history of the Moon might illuminate.

At the stage of Apollo 14, the nature of the Moon itself and its ancient relation to the Earth eluded them, just as the nature of the North and South American continents and their relationship to Europe and Asia eluded Columbus, Cabot, and the mariners who sailed later in the sixteenth century. Throughout most of that century, captains and mapmakers believed that Newfoundland, which Cabot rediscovered 500 years after Leif Ericson built an ill-fated settlement there, was a peninsula of Asia. They believed that beyond it, to the south and west, lay a short water route to China and Japan.

In many respects, the Apollo scientific teams rediscovered what Gilbert had surmised by purely visual observation through a telescope in the nineteenth century. But even so, the scientific community appeared in 1970 to be as divided and confused about the nature of the Moon and its origin as the Europeans of the sixteenth century were about the lands whose shores they had barely touched.

The search for a Northwest Passage dominated the explorations of Joao Alvares Fagundes, sailing for Portugal, and Jacques Cartier and Giovanni Verrazzano, sailing for France. It was the main objective of the Elizabethan captains, Martin Frobisher and John Davis. To these people, the Gulf of St. Lawrence, Davis Strait, and Frobisher Bay looked like possible passages to China. Verrazzano believed that the Carolina banks represented an isthmus between the Atlantic and the Eastern Sea, an idea perpetuated by Italian cartographers from his reports for most of the sixteenth century.[2] In this period the discovery of a continental landmass worth colonizing was subordinate to the search for sea routes and gold. For 150 years Europeans sought riches in fabled regions, such as the mythical land of Norumbega, an Elysian paradise Indians along the Penobscot River alluded to in the hope of encouraging European marauders to move on.

Nearly every period of exploration has its Norumbega, its Seven Cities of Cibola, and its Hy Brasil. This is the chimerical phase of exploration, dominated by myth and speculation. In the eighteenth and early nineteenth centuries, English, French, and Russian expeditions sought the fabled "Southern Continent," a land deemed to be as fair and rich as any yet discovered, lying somewhere in the cold and stormy Southern Ocean.

In a like manner, the scientists of Apollo believed the Moon would reveal to them the secrets of the Solar System and of the creation of the planets. They thought of it as a laboratory, as Antarctica had become during and after the International Geophysical Year of 1957–1958, from which Solar System cosmogony could be studied and ultimately understood. The scientific motive was not too far removed from the patent avarice that launched the overseas forays four hundred years before. It was simply more sophisticated. The industrial revolution of the nineteenth century and the scientific revolution of the mid-twentieth century had shown that research is the key to science, that science is the key to technology, and that technology is the key to wealth, power, and survival in a populous society.

The Atlantic voyages of the sixteenth century had another point in common with the lunar voyages of the twentieth century in their indifference to the prospect of colonization. Unless minerals were found, there seemed to be little to exploit on the coasts of Newfoundland and Labrador, even in the wilderness of New York Bay. There was nothing on the Moon to exploit, not even water, which could be a source of oxygen and (hydrogen) fuel for colonists. The idea of colonizing the Moon, except for a short-term science base, was too far fetched to contemplate. Europeans of the sixteenth century regarded North America in the same way. Like the Moon, it was a fascinating place to visit, but who would want to live there?

The Moon of 1970

How much did we really know about the Moon after two voyages? Scientists in the lunar program summarized their knowledge and speculation in pre-Apollo 14 flight conferences with the news media at the Manned Spacecraft Center, Houston, and at the Kennedy Space Center, Florida.

At these meetings the scientists spoke more freely and much less self-consciously than at formal scientific sessions. Because of this informality, they felt subject to fewer constraints and were more inclined to speculate about the significance of their findings.

Moreover, because they were talking mostly to generalist reporters and writers with only a smattering of education, if any, in the earth sciences, physics, chemistry, and mathematics, the Apollo scientists had to translate the jargon in which they communicated among themselves into its ordinary English equivalents. They had to translate scientific ideas and nomenclature for most of the mass media representatives if they hoped to convey the significance of their costly experiments to the taxpayers. There were, of course, a group of experts among reporters, writers, and broadcasters, people who had been writing science for years and who had done their homework in the Apollo program. But they were a minority. With a few exceptions, much of the scientific detail was omitted in flight reportage. It took nearly all the space that reporters had to tell the logistics of an Apollo flight—the liftoff, the journey to the Moon, the problems on the surface, the return, and the splashdown. What the astronaut heroes were really trying to do up there and how well they were succeeding in unravelling the mysteries of the universe hardly ever was told. There simply was not enough space in the newspapers nor time on the air to tell it.

The scientific achievements of the lunar adventure were reported mainly in the scientific journals, but in such a framented and cryptic way that only experts could recognize their significance. The general public, which was financing the lunar ballgame, had little opportunity to learn the score. The consequences were an increase in political hostility toward Apollo and a loss of public interest in it.

An Emerging Hypothesis

The science briefings and press conferences, therefore, form an historical record of how scientists themselves developed their ideas about the Moon from one mission to another.

In reviewing hundreds of pages of the transcripts of these meetings, as well as my own notes and tape recordings, I am impressed by the intensity with which these specialists tried to communicate their ideas. Some were excellent teachers. Few were hesitant about speculating on the implications of their findings. For the most part, they were established well enough in their own fields to risk being wrong.

The initial Apollo 14 Pre-Mission Science Briefing was held in Houston, January 8, 1971. Wood, of the Smithsonian Astrophysical Observatory, made a comment that summarized an important distinction between the Earth and the Moon and that later was hotly disputed.

Wood noted that so far, the astronauts had failed to find bedrock on the Moon. All was rubble and dust that was compacted at depth. No one had encountered any extensive layer of unbroken hard rock under the topsoil or regolith.

"It's from this layer of broken-up rubbish that the astronauts make their collection," Wood said. The process that formed this outer part was understood. When meteorites or asteroids or comets struck the lunar surface, several things happened. A small amount of the topsoil or rock was melted and vaporized. Melted material ejected from the crater condensed into little droplets of glass. A larger amount underneath was compacted into solid material—the aggregate type of rock called breccia. Then, a much larger quantity was broken up, but not changed very much. This mechanism explained the breccia and the regolith as products of bombardment.

Then there were the dark igneous rocks that were broken-up pieces of lava. This was the dark material of the lunar seas that had formed a billion to a billion and a half years after the Moon is believed to have accreted. The lava was basalt in which the principal mineral was ilmenite, a titanium compound that is black and gives the lavas of the maria their dark appearance. This can be seen quite readily from Earth with the naked eye. There were also a white feldspar and a beige pyroxene.

So much seemed to have been established. However, Wood said, something quite different from these materials underlay and surrounded the basalt. It formed the receptacle into which the basalt began flowing a billion years after the genesis of the Moon.

Was this the anorthosite crust that he and his colleagues had hypothesized from the Apollo 11 fragments? Perhaps the journey to Fra

Mauro would tell. Or would Fra Mauro turn out to be another Northwest Passage? Another Norumbega?

Gary Latham, spokesman for the seismology team, offered an optimistic view. On the eve of the Fra Mauro expedition, it appeared to him that "most geologists and geophysicists would agree that a good working hypothesis for the Moon was emerging."[3]

There was the molten outer shell, which Wood had described at the Apollo 11 Scientific Conference. It covered the whole Moon. Then the missiles struck, gouging out the mare basins, such as Imbrium. The accumulation of heat caused by the decay of radioactive minerals in the soil (potassium, thorium, and uranium) had melted the material below the surface at depths of perhaps 100 miles. The lavas welled up, as Wood had described, and flowed into the basins to form these dark seas. Since that happened, Latham surmised, there had been little activity except sculpturing by meteorite and meteoroid impacts. Perhaps the interior of the Moon, down to the center, never was molten. In the outer regions, however, there was "pretty good evidence" that it was, Latham said.

There was widespread but by no means complete agreement on this picture. But more confirmation was needed. The question confronting the seismology team was, where was the base of this outer, crustal shell? On the Earth, the sound vibrations produced by earthquakes travel through the crust at one speed and through the mantle much faster. The border between the crust and mantle is called the Mohorovičić discontinuity (or Moho) after the Yugoslav seismologist Andrija Mohorovičić, who first noticed it.

Was there a Moho on the Moon? A boundary between crust and mantle? Or was the Moon simply an undifferentiated lump of rubbly material to the center?

So far, there was only one seismic instrument on the Moon that was working—the seismometer at the Apollo 12 site in Oceanus Procellarum. The impact of the Apollo 13 S4B had produced vibrations in the Moon that indicated broken-up, unconsolidated material for a depth of 20 miles.

Perhaps the impact of the Apollo 14 S4B rocket stage would extend this picture to a depth of 50 miles. Then, too, with a successful landing, another seismometer array would be emplaced at a strategic location.

Trying to figure out the interior of the Moon with only a single instrument was quite frustrating, Latham said. He explained that on the Earth there were a thousand seismic stations, and this number was not regarded as too high by any means. "If you gave the average seismologist one station somewhere in a continent on Earth, he would either quit or go mad in the process of trying to analyze the data," Latham said.

To approximate the results of the terrestrial seismic network about 30 stations would be required on the near side of the Moon alone.

Some data had been accumulated from the Apollo 12 seismometer. During the year it had been operating, moonquakes and meteoroid impacts had been recorded on the average of a little less than once a day. Thanks to the LM ascent stage and S4B impacts, the seismic team had learned to distinguish a quake signal from an impact signal on the seismogram at Houston, to which the radio signals from the instrument on the Moon were fed.

Moonquakes seemed to be concentrated in one place—near a well-developed set of rilles (canyons) 80 to 100 miles south of the Fra Mauro Crater, the feature after which the Fra Mauro formation was named. But the exact location was not pinned down.

As I have mentioned earlier, the number of quakes increased near perigee, the point in the Moon's orbit where it is nearest the Earth. Latham attributed the increase to tidal stress, which caused the surface to bulge out toward the Earth as much as 20 or 30 inches.

Although tidal stress might be the trigger for these quakes, Latham believed there was another source of energy that produced them in concert with the tide on the Moon. It must be a molten region deep within the Moon, but more data were needed to confirm it.

Meteorite impacts were infrequent. Within a 200-mile radius of the station, the impact of an object the size of a grapefruit was picked up on the average about once a month. There was less debris in space near the Earth–Moon system than expected from Earth-based measurements, Latham said. The chances of an astronaut's getting beaned on the Moon were no greater than his chances of getting beaned in Texas. "So," Latham added, "the job of building a permanent or semipermanent base on the lunar surface is not going to be as tough from the point of view of meteoroid hazards as might have been guessed earlier."

Although no one was planning at that time to build a scientific base on the Moon, the idea had been considered by a presidential Space Task Group in 1969. A number of geologists believed that a full understanding of the Moon required a full-scale geological survey lasting many years, perhaps a century. More than one base would be required for that—and so would a transportation technology less costly than Apollo and capable of carrying bigger payloads.

Shepard Flies Again

The Apollo 14 astronauts assigned to explore the Fra Mauro landing site were Navy Captain Alan B. Shepard, Jr., 47, and Navy Commander Edgar Dean Mitchell, 41. The youngest crewman was Air Force Major Stuart A. Roosa, 37. As command module pilot, he would remain aboard the Apollo spacecraft, whose call sign was Kitty Hawk, while Shepard and Mitchell descended to surface in the lunar module Antares.

Shepard is remembered as the first American in space. He was one of the survivors (in the program) of the original Mercury Seven group of astronauts, appointed in the spring of 1959 as Project Mercury was getting under way. On May 5, 1961, the 37-year-old Navy Commander climbed into the one-man capsule, Freedom 7, which sat atop a Redstone rocket on Cape Canaveral. And while 600 watchers screamed, shouted, applauded, and wept, he was rocketed aloft to an altitude of 116 miles on a 15-minute ballistic flight. The tiny capsule parachuted gently into the Atlantic 302 miles southeast of the cape along what was then known as the Atlantic Missile Range (later the Eastern Test Range).

Although the Russian cosmonaut, Major Yuri Gagarin, had upstaged the flight of Shepard's Mercury-Redstone 3 the previous April 12 with a 108-minute flight around the world in Vostok (East) I, Shepard instantly became an American hero in the Charles A. Lindbergh tradition.

Handsome and poised, Shepard remained a memorable figure even after John H. Glenn, Jr. rode the Mercury capsule, Friendship 7, three times around the world February 20, 1962. But Shepard did not fly again in Mercury, Project Gemini, and the early stages of Apollo. He became an executive in the program and prospered in outside business ventures as the years passed. Now, after nearly 10 years, he was going to the Moon, a prospect that had seemed fantastic in 1961. On January 9, 1971, he met with reporters and was asked how he felt to be going all the way at last.

"What can you say about 10 years of fantastic technical progress?" he said. "I think it's something of which I've been very proud to be a part. I'm happy to see that we've made as much progress as we have. I'm unhappy that we haven't made more."

A great deal had changed in that decade. The Russians had dropped out of the manned race to the Moon. Following an abortive attempt in July 1969 to return a lunar soil sample to Earth via a robot lander (Luna 15), the Russians successfully scooped up a quarter of a pound of soil on September 20, 1970, in the Sea of Fertility with the automatic craft, Luna 16, which returned it to Earth in the Soviet Kazakh Republic on September 24. Some of it was exchanged with American scientists in return for Apollo samples. This was the first effort to internationalize the lunar analysis program, although data had been exchanged at international scientific meetings since the start of the space age. In November 1970, the Russians guided another automatic craft, Luna 17, to a landing in Mare Imbrium, and out crawled the 1600-pound, eight-wheeled roving vehicle called Lunokhod (Moon Rover) I. By the time Apollo 14 was ready for launching, Lunokhod had made 200 soil measurements and a televised topographic survey of a strip of lunar ground 2 miles long and 500 feet wide. It had provided a base for precise laser ranging measurements as well.

A New Hampshire man and a graduate of the U.S. Naval Academy at Annapolis, Shepard began his military career during World War II aboard a destroyer in the Pacific Ocean and then entered flight training. He was graduated from the Navy Test Pilot School at Patuxent, Maryland. He married the former Louise Brewer of Kennett Square, Pennsylvania. They had two daughters, Mrs. Laura Shepard Snyder, 23, and Julia, 19.

In 1963 Shepard began experiencing the symptoms of an inner ear disorder called Menière's disease. It produced vertigo, buzzing, and some hearing loss, and this caused him to be grounded after treatment failed to relieve the condition.

During this period, when Gemini spacecraft were flying, Shepard was Chief of the Astronaut Office at the Manned Spacecraft Center, Houston. He became a key figure in the assignment and training of Gemini and early Apollo crews under the supervision of another Mercury astronaut who was grounded by a heart rhythm irregularity, Donald K. (Deke) Slayton. Slayton had been appointed Director of Flight Crew Operations in November 1963.

The two presented a firm and sometimes arbitrary management image to the corps to which 73 test pilots and scientists were appointed between 1959 and 1973. Shepard and Slayton were the men who largely determined flight crew selection policy that gave flight priority to test pilots. This policy kept scientist astronauts grounded so long that many of them resigned. As a result of it, only one scientist astronaut ever went to the Moon—but all the crews came back safely.

Slayton believed in crewing Apollo vessels with experienced test pilots on the theory that these vehicles were still experimental and that pilots were more likely to react effectively to emergencies than scientists with only the jet pilot training they received from NASA.

Shepard backed him up. So did the Manned Spacecraft Center management. Robert R. Gilruth, the director, and his chief deputy, Christopher Columbus Kraft, were aeronautical engineers and wore the same school tie as the test pilots. They tended to regard the one-man Mercury, two-man Gemini, and three-man Apollo spacecraft as developmental vehicles requiring experienced test pilots. NASA headquarters in Washington supported this policy. Headquarters would not risk a pilot-error accident and thus underwrote the policy that allowed only the most experienced pilots to fly to the Moon until the final voyage, Apollo 17.

Shepard was determined to be one of this elite group. Although a Christian Scientist, he underwent surgery in 1968 to correct the Menière syndrome, which was caused by a build-up of fluid pressure in the inner ear. The operation relieved all the symptoms. After healing was complete, Shepard was retested and restored to flight status in May 1969.

Among the astronauts, Shepard was regarded as the most successful in business matters and was reputed to have become well-to-do. At his January 9 press conference, he was asked: "Captain Shepard, you've had some success in the business world . . . why don't you simply choose to stay in your mansion in River Oaks instead of going to the Moon? What's kept you going at it this long?"

"Well," said Shepard, "I've had some failures in the business world, also. With respect to your question about why I want to continue to fly in space, I guess it's because it's about the only business that I know. I've been trained as an aviator for years. Really trained as nothing else. It's something to which I enjoy making a contribution. It's something I believe in and fortunately I'm healthy enough to be able to get another chance to go."

Shepard's Moon partner, Mitchell, a Navy Commander, was born in Texas and had worked for a while as a cowboy in Arizona. He could ride, rope, and brand cattle as well as fly jet aircraft. He had been fascinated by aviation since his boyhood, most of which was spent in New Mexico. In spite of his early cowpunching background, Mitchell was regarded as one of the more academic astronauts. He was intrigued by research into extrasensory perception, which he regarded as a fascinating scientific problem.

He held degrees in industrial management from the Carnegie Institute of Technology, in aeronautical engineering from the Naval Postgraduate School, and a Doctor of Science degree in Aeronautics and Astronautics from the Massachusetts Institute of Technology. After joining the Navy in 1952, he became a research pilot with an air development squadron. Prior to his appointment as an astronaut in 1966, Mitchell had been Chief of the Project Management Division of the Navy Field Office for the Manned Orbiting Laboratory. This was a Department of Defense project that was abandoned in favor of NASA's less expensive Skylab program.

Mitchell married the former Louise Elizabeth Randall of Muskegon,

Michigan, and they had two daughters, Karlyn, 17, and Elizabeth, 11.

Neither he nor Stuart Roosa had been in space before, and Shepard was ahead of them by only 15 minutes. The crew, although seasoned jet pilots, were relatively inexperienced in actual space flight. Roosa, an Oklahoman with a Bachelor of Science degree in aeronautical engineering from the University of Colorado, had been on active duty with the Air Force since 1953. He was a graduate of the Aerospace Research Pilots School, and his last assignment before joining NASA in 1966 was as experimental test pilot at Edwards Air Force Base, California.

A fan of country and western music, Roosa brought some of his favorite tapes to play while he was cruising around the Moon in orbit, waiting for Shepard and Mitchell to complete their work on the surface. He called himself a "hillbilly at heart" and had acquired the nickname "Smoky" as a result of working in the Forest Service as a smoke jumper. He married the former Joan C. Barrett of Tupelo, Mississippi. Their children were Christopher, 11; John, 10; Stuart, Jr., 8; and Rosemary, 7.

Although exceptionally well schooled to fly the lunar mission, the crew had acquired only superficial training in the Earth sciences. They had received general classroom instruction in geology and had taken eight field trips to volcanic regions in Hawaii, Mexico, and the Southwest. They had even examined an ancient impact structure in Germany.

At the Atomic Energy Commission's Nevada Test Site they inspected craters made by tests of nuclear explosives. One crater, excavated by a test shot called Schooner, was believed to be comparable in size and structure to a lunar feature called Cone Crater at the Fra Mauro landing site. Approximately 1105 feet across and 250 feet deep, Cone Crater was a prime objective, for mission scientists were convinced that the ancient rocks they were seeking to analyze would be found there.

The Bubble Men from Jupiter

Apollo 14 was launched at 4:03 P.M., EST, January 31, 1971, from the Kennedy Space Center, about 40 minutes late because of heavy clouds. The delay made it necessary to alter the navigation plan in order to land on target. During the flight, the crew encountered two problems, either of which could have wrecked the mission if not solved. The first was exceptional difficulty in docking the Apollo command and service module with the lunar module in the S4B stage during the transposition and docking phase of the journey. Shepard finally succeeded in docking on the sixth attempt.

Early February 4, 1971, after Apollo 14 was inserted into lunar orbit, problem number two showed up. The Abort Guidance System in the lunar module Antares revealed a faulty switch, and this caused the system to abort the landing program. With some coaching from Mission Control, Shepard and Mitchell devised a landing routine that would bypass the defective switch.

Antares touched down in a broad, shallow valley at 2:37 A.M., February 5, 1971, as Shepard and Mitchell stood at the controls like turn-of-the-century streetcar motormen. They landed 160 feet from the targeted landing point at latitude 3 degrees, 40 minutes, and 24 seconds south and longitude 17 degrees, 27 minutes, and 55 seconds west. They were 738 miles south of the center of Mare Imbrium, 216 miles south of the Crater Copernicus, from which ejecta was strewn over the landing area, and 110 miles east of the Apollo 12 station.

Antares stood on a slope, tilted at 7 degrees to starboard, but this did not affect its stability for the launching of the ascent stage later.

"We're here," Shepard informed Mission Control noncommitally. Looking out the window, he and Mitchell concluded that they had come down in a bowl. Two small craters, marked on their maps as Doublet Crater, seemed to be 25 to 30 feet above them. The terrain, said Shepard, "is a little rougher than I expected."

Mitchell could see several ridges and rolling hills 35 to 40 feet high. Below, the regolith looked brown and gray. They quickly ate meals of salmon salad, beef and gravy, jellied candy, and a grape drink and prepared for their first EVA (extra-vehicular activity). The first task was to collect samples of rocks and soils near the landing site. Then they would set up the ALSEP (Apollo lunar scientific experiment package) instruments under the eye of a color television camera and execute a seismic experiment that would probe the depth of the regolith.

Shepard climbed down the ladder at 8:50 A.M., CST, at Houston, where control of the flight had shifted immediately after the launch from the Kennedy Space Center. It was bright in the Friday morning sunrise. Shepard remarked that Cone Crater, their objective on their second EVA, was "right where it should be—a very impressive sight." Mitchell then descended to the surface and Shepard collected a contingency sample about 25 feet from the LM. He then set up the television camera on a tripod about 100 feet away. He was careful to keep the lens away from the Sun, which had blinded the Apollo 12 camera. Now, for the first time, there would be a televised record of man on the Moon.

The two moved about with mincing, dancing steps, which soon escalated to kangaroo jumps as they became accustomed to the one-sixth gravity of the Moon. They set up an S-band antenna, which looked like

APOLLO ⑭ TRAVERSE

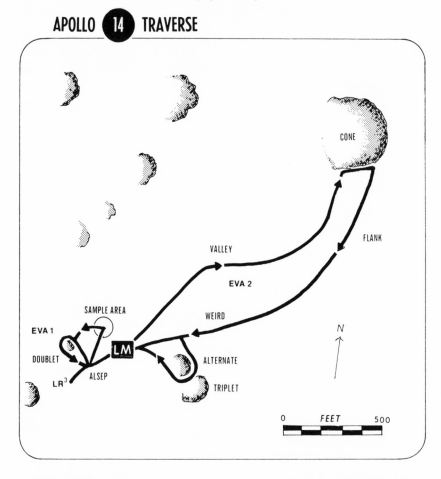

CONE

FLANK

VALLEY

EVA 2

SAMPLE AREA

EVA 1

WEIRD

N

DOUBLET

LM

ALTERNATE

LR³   ALSEP

TRIPLET

0   FEET   500

an inside-out umbrella, and pointed it toward Earth, which hung just above the south horizon.

In the television pictures that came to Earth from 238,000 miles away, the explorers looked like bulky white ghosts against a black sky, cavorting about a strange landscape of dunes and craters that were glaring white where the sunlight fell in them and were jet black in shadows.

Someone in Mission Control remarked that Shepard and Mitchell looked like the "Bubble Men from Jupiter" on one of those Saturday-morning TV shows for children.

Unpacking the ALSEP from the Antares' MESA (modularized equipment stowage assembly) on the descent stage, Shepard and Mitchell deployed two seismic experiments, a passive one consisting of seismometers that sensed Moonquakes and impacts, like the instrument at the Apollo 12 site, and an active one in which they participated.

They set up also the solar wind collector, the ionosphere and atmosphere detectors, which had been carried also on Apollo 12, and a spectrometer designed to measure the energy range of protons and electrons in the solar wind. The aluminum-foil solar wind particle collector, mounted on a staff, simply trapped the nuclei of elements in the solar wind while the spectrometer measured the charged particle energies.

They also erected the second Apollo laser mirror to be installed on the lunar surface. The first one had been set up by Armstrong and Aldrin at Tranquility Base. Eventually, a network of three U.S. laser reflectors would be emplaced on the front side of the Moon. By reflecting laser beams from Earth observatories they would provide information on the Moon's distance from the Earth at various points in its orbit to within 15 centimeters, and, over a long period of time, they would confirm a theory that the Moon is receding from the Earth at 3 centimeters a year. Laser measurements also would show fluctuations in the Earth's rate of rotation, the amount of wobble in the Earth's axis, a phenomenon that possibly influenced climate, and any change in the gravitation constant, *G*, which some theorists believed to be dwindling.

The Thumper

The active seismic experiment relied on man-made seismic waves to analyze the structure of the Moon. The waves, like sound waves that travel through rock instead of air or water, were to be produced by explosions. They would be "heard" or picked up by instruments called geophones, which were essentially small seismometers, laid out on the ground at some distance away from the explosions. It was this kind of seismology that first revealed the true depth of the antarctic ice cap during the International Geophysical Year, when teams of geologists set out strings of geophones on the ice and then exploded dynamite in shallow holes to produce the sound waves. The time it took those waves to hit rock bottom and be reflected back to the geophones on the surface of the ice sheet gave the ice depth. The ice sheet turned out to be much thicker than anyone had surmised, and scientists the world over had to revise their estimates of the Earth's water budget.

On the Moon, the problem was somewhat different. The experimenters wanted to know the depth of the rubbly regolith, at what depth there might be bedrock, and what the structure might be below it. The velocity of the seismic waves as measured by the geophones would reveal the nature of the material through which they were passing. In general, these waves traveled faster through dense material than through broken-up rubble or loose material. The three geophones in the active seismic experiment converted the seismic waves into electrical signals, which were radioed to Houston.

During the first EVA, which lasted 4 hours and 49 minutes, Shepard and Mitchell set up the geophones at intervals of 10, 160, and 310 feet from the ALSEP central power and transmitter station. When this was done, and the geophones were checked by Houston, Shepard and Mitchell were given a "Go" for "thumper activity."

The thumper was one part of the experiment that generated seismic waves. It was a staff 44½ inches long containing 21 explosive charges each with the force of a Fourth of July "cherry bomb." Mitchell carried the staff and was supposed to fire it at 15-foot intervals along the geophone line.

For the initial thump, Mission Control staged a countdown. "Five, four, three, two, one, fire!" Mitchell pulled the trigger in the thumper handle to detonate the first shot but nothing happened. He paused a moment and tried again. This time the thumper thumped. Scientists at Houston were delighted with the recording. Mitchell continued to fire the thumper, but occasionally it misfired. All told, he succeeded in getting 13 detonations from the 21 charges.

When this experiment was completed, Shepard collected about 8 pounds of rocks and soil near Antares under the verbal direction of Paul Gast, who had joined the Apollo scientific staff as chief scientist. Gast could see Shepard's blimplike white figure oscillating about in the TV screen. Never before had exploration been masterminded from such a distance. As far as the scientists at Houston were concerned, the astronauts on the Moon were simply an extension of eyes, hands, and feet of the principal investigators. It was not hard to imagine explorers working on the surface of Mars under televised observation and radio direction from Houston some 40 million miles away.

Shepard then "uncaged" the passive seismometers that had been secured during thumper activity. At once, the seismometers designed to detect distant impacts and weak Moonquakes picked up the footfalls of the kangarooing Moon men. This activity was transmitted to Houston as a faint warbling sound that accompanied the pen squiggles on the seismogram. The receivers at Houston had been equipped with an audio system to convert seismometer signals into analogous sound frequencies proportional in frequency to those generated on the Moon. The warbling sounded like the old Rudy Vallee number, "Tiptoe Through the Tulips." But controllers referred to the lunar warbling as "looney tunes."

In addition to the thumper, there was another explosive component to the active seismic experiment. It was a battery of four rocket grenades containing 0.10 to 1 pound of high explosives. The rocket grenades were to be launched by radio command from Houston from a mortar box that Shepard and Mitchell set up at the site—a little too close, as it turned out, to the Central Station.

Months after the astronauts had left the Moon the rockets were to be fired by remote control. They were fueled to carry the grenades to distances of 500, 1000, 3000, and 5000 feet where the grenades would explode to generate seismic energy.

It had been a long and busy EVA and the Moon men went about their tasks like ambitious workers. They did everything they were supposed to do, and they did it right. Houston was pleased, even though everyone had been up all night.

The Second EVA

Meanwhile, the Apollo 14 S4B, weighing 30,836 pounds and traveling 5,683 miles an hour, crashed 103.5 miles southwest of the Apollo 12 site at 1:40 A.M., February 4, while the entire crew was still in orbit. Houston said the energy of the impact equaled that of 11 tons of TNT. Like the Apollo 13 S4B crash, it produced a long seismic signal that reverberated for 3 hours.

Preliminary analysis showed that there was no appreciable change in the composition of the surface to a depth of 21 to 24 miles. On Earth the mantle would have been seen at that depth in ocean beds, but not on the Moon. Either the Moon had a much thicker crust than the Earth has, or it had no crust at all, at least in the Oceanus Procellarum region.

Coincidentally, the Russian News Agency Tass quoted a Professor Pyotr Kropotkin as saying that the lunar crust beneath the maria was only 1.2 miles thick, although elsewhere the thickness ranged up to 36 miles.[4] Professor Kropotkin's estimates were derived from a gravimetric map of the Moon. But a satisfactory resolution of the problem of a lunar crust was not to come yet.

Several hours after EVA 1 was concluded, a preliminary interpretation of the active seismic experiment from thumper data indicated a layered structure as far as the 100-foot depth to which thumper sounds had penetrated. Some of the explosion energy had been reflected, indicating a denser layer 28 feet below the surface. The seismic waves moved at 340 feet per second through the 28-foot layer, which indicated powdery, unconsolidated soil. Below 28 feet the compressional (P) wave velocity rose to 980 feet per second, indicating denser material. An Earth analogy would be alluvial soil over bedrock.[5]

Early Saturday, February 6, the inmates of Antares climbed out of their hammocks after a fitful sleep and squinted out the window at the lunar desert. It was still morning on the Moon, dazzling bright in the sunshine but pitch dark in the shade. Today they were to break trail to Cone Crater, their main objective. The 340-meter (1105-foot) crater had been excavated by some ancient impact on a ridge of an Imbrium throw-out, the ejecta that had flowed radially outward from the Basin. The ejecta itself had been covered substantially by regolith. Lunar Orbiter photos showed that the bottom of Cone Crater went down well below the regolith into a stratum of huge blocks or bedrock. This stratum, it was expected, was Imbrium ejecta and probably consisted of early crustal rocks. Some of these blocks had been splashed out of the bottom of Cone Crater and now stood massively on the rim—within reach of Shepard and Mitchell. That was the whole strategy of landing at this site, and, as they began EVA 2, the explorers were within walking distance of this goal, thanks to their pinpoint landing.

Wheels on the Moon

The task confronting Shepard and Mitchell now was to march up to these blocks on the crater's rim, chip off some pieces with a hammer, photograph the crater's interior (estimated to be 250 feet deep), take magnetic readings with a portable magnetometer, take extensive panoramic photos and documentation shots of the samples they would collect, return to Antares in time to clean up the site, and take off to rejoin Roosa in Kitty Hawk at the appointed hour. To do all this, Mission Control allowed them 4 hours, later extended by 20 minutes. It was a tight schedule.

The Apollo directorate had supplied this expedition with a hand-drawn cart that had two rubber-tired wheels and suggested an oriental rickshaw. It was called in typical aerospace style a "modularized equipment transporter" or MET. It enabled Shepard and Mitchell to carry a good deal more equipment and rock samples than they could have done by hand. But of significance to later missions, the MET was a test of how a wheeled vehicle would operate on the Moon. Would it become mired in the dust of the regolith? How stable must it be to avoid overturning when one wheel or the other encountered potholes? This information was essential for the operation of the four-wheel electric cars that would be carried to the Moon on the last three missions.

The journey to Cone Crater was barely 3 kilometers (1.8 miles), but the outward-bound leg of the traverse was mostly uphill. Shepard and

Mitchell took turns pulling the MET. It seemed to roll easily over the soft soil on its rubber tires.

Soon, however, the explorers were puffing and their heart rates were climbing. Although they labored under only one-sixth of the gravity they were used to, they had to fight their semirigid, cumbersome suits. With their life-support systems on their backs, they were carrying a sizable fraction of their Earth weight. Moreover, they were working under muscular tension and timeline pressure. Wariness about the strange, hostile environment in which a fall could be fatal added to the pressure. Despite the tension and the uncertainty about what might lie beyond the next ridge, however, they felt a strong sense of exhilaration at the strange, barren beauty of the lunarscape.

As they climbed the slopes of the big ridge in which Cone Crater had been punched, they heard the voice of Fred Haise, who had missed his chance on the Moon in the Apollo 13 debacle, calling from Mission Control. He wanted them to describe the soil.

"The surface here is textured, a very fine, grained dust," Shepard answered. "About the same as in the vicinity of the LM. But there seems to be more small pebbles here. Also more boulder-sized rocks." Mitchell replied, "Y'all might comment, Fred-O, that they have the appearance of raindrops, of a raindrop-splattered surface."

About 150 meters from Antares, they reached "Point A" on their maps. Shepard took a double core sample (36 inches deep) by pounding the hollow core tube into the regolith. Mitchell set up the portable magnetometer and recited its readings to Houston.

The instrument detected a magnetic field here that was 3 times larger than the field found at the Apollo 12 site. No one had expected that. It would reinforce the theory that the Moon had an iron core and had once been molten.

After picking up a rock with the tongs—it was hard to bend over in the Moon suits and the tongs were standard rock-collecting equipment —they pressed on. Now they were moving downgrade and the going was easier. Mitchell reported that the cart tracks were smooth, like those of a tractor in a plowed field. The plowed field analogy had struck everyone who examined the regolith since the first Surveyor photos were transmitted to Earth. These lunar fields had been plowed by cosmic forces, by the rain of planetesimals, asteroids, comets, secondary fragments heaved out of big craters, meteoroids, and micrometeoroids.

Shepard and Mitchell crossed a shallow valley and once more resumed the climb up to the ridge. Boulders 4 and 5 feet in diameter were beginning to appear. But the landscape was becoming more hilly, and it was difficult to follow the map because the profusion of ridges and hollows hid key landmarks. To the south of their track was a landmark called Weird Crater because of its elliptical shape, but it was not visible. They had trouble finding "Point B" on the map. They were supposed to pick up rock samples there. Fred Haise advised that the general area of "Point B" would be good enough. They were running 15 minutes behind their time line now and they still had a long way to go.

The Elusive Rim

Presently, the south flank of Cone Crater came into view. It was the largest feature on their horizon. The ground was undulating and pocked with ancient craters that were hardly 10 yards apart, most of them worn down by cosmic erosion.

SHEPARD: The ridge of Cone Crater to the north is now very apparent as we expected it would be. It stretches off into the distance and meets with the far horizon.

MITCHELL: We're starting uphill now, fairly gentle at this point, but still a hill.

Shepard dropped down on one knee to pick up a rock. He struggled to regain his feet and Mitchell helped him. The difficulty of just rising from one knee underscored the danger of a fall, especially on a slope.

As they climbed, Shepard reported seeing numerous fresh craters. They seemed to be hardly eroded at all and were spaced 75 to 100 feet apart. Some were 25 to 30 feet wide and 5 to 6 feet deep. In one lay a large rock, covered with glass.

The explorers adjusted their suit-cooling systems from low to medium. They were working hard and the Sun was heating up the lunar desert. Haise advised a rest. They were about two-thirds of the way to Cone Crater, and they halted, studying the map.

Looking back toward the west, Mitchell spied Antares in the distance, leaning toward the south.

MITCHELL: That old LM looks like it's got a flat tire.

HAISE: Say again, Ed?

MITCHELL: Just talking. Never mind.

At this point in the traverse, Shepard and Mitchell were showing heart rates of 120 beats per minute while moving uphill. This was only medium high for them.

The grade became steeper and the soil firmer as they ascended the ridge. They photographed a boulder 12 feet long and 4 feet wide. Other big rocks loomed up ahead, like silent sentinels. Most of them were taller than the astronauts, each of whom was 5 feet 11 inches.

This was big rock country, and Shepard became convinced they had finally reached the blocks that they had come all this way to sample. He reported some of the rocks showed "weathering" like Earth rocks, although from a different process. This weathering seemed to be a universal effect.

They rested and their heavy breathing could be heard in their radio transmissions to Houston. The sound of men breathing hard on the Moon was heard in Houston, in New York and Washington, and in Florida. It symbolized man's struggle to explore the cosmos. Mitchell and Shepard began to discuss their next moves.

MITCHELL: Well, we're three-fourths there. Why don't we lose our bet, Al, and leave the MET and get on up there? We could make it faster.

He referred to a bet they had made with the backup crew, who wagered they would dump the cart when the going got tough.

SHEPARD: No. I think that what we're looking at right here, this boulder field, Ed, is the stuff that's ejected from Cone.

MITCHELL: But not the lowermost part, which is what we're interested in.

SHEPARD: Okay. We'll press on a little further, Houston. And keep your eye on the time.

HAISE: Okay. And as of right now, we have a 30-minute extension to the EVA time. We'll stand by.

Shepard paused to take a panoramic photograph from the ridge. The view was fantastic, he said. They pushed on toward the crater's west rim. From Houston came words of encouragement.

HAISE: Al and Ed. Deke [Slayton] says he'll cover the bet if you drop the MET.

MITCHELL: We need those tools.

SHEPARD: No, the MET's not slowing us down, Houston. It's just a question of time. We'll get there.

The going became more and more difficult. It was easier occasionally to pick up the MET and carry it, than haul it uphill. As they struggled upward, the rim of the crater seemed to be just ahead, but they never seemed to get nearer to it. They rested a while.

MITCHELL (to Houston): You can sure be deceived by the slopes here. The Sun angle is very deceiving.

(To Shepard): Okay, let me pull [the cart] a while. You ready to go?

SHEPARD: Houston, we're proceeding now.

MITCHELL: These rocks and boulders are getting more numerous toward the top here. However, it's nothing like the rubble of large boulders that we saw at the Nevada Test Site.

They climbed and so did their heart rates, Shepard's to 150 beats per minute and Mitchell's to 128. Take another break, Haise counseled.

Scanning the skyline and then the map, Shepard estimated that the elusive rim of Cone Crater was still 30 minutes away. They had spent more than half their 4-hour EVA time getting this far. Even with the 30-minute extension, they would have to hurry to return to Antares on schedule. Shepard concluded there was not enough time left in the traverse to reach the rim.

MITCHELL: Oh, let's give it a whirl. Gee whiz! We can't stop without looking into Cone Crater.

SHEPARD: I think we'll waste an awful lot of time traveling and not too much documenting.

MITCHELL: Well, the information we're going to find, I think, is going to be right on top. Fred-O, how far behind the time line are we?

HAISE: The best I can tell right now—about 25 minutes down. The word from the back room is—they'd like you to consider where you are at the edge of Cone Crater. That decision, I guess, was based on Al's estimate of another 30 minutes. Do you have the rim in sight at this time?

MITCHELL: Oh, yes.

SHEPARD: It's affirmative. It's down in the valley.

[They thought Haise was asking about the LM.]

HAISE: Sorry. You misunderstood the question. I meant the rim of Cone Crater.

SHEPARD: Oh, the rim! That is negative. We don't . . . we haven't found that yet.

Haise advised them they had exceeded their 30-minute extension and asked them to collect samples. They took another magnetometer reading, which again showed a field higher than that at the Apollo 12 site. When they attempted to get a core sample, the grainy material fell out of the core and Shepard gathered it up and put it in a bag. He noted that below the brownish topsoil, the undersoil was whitish.

Mitchell reported a rock 25 feet long. Broken boulders that were brown on the outside showed white inside. Judging from the experience of Conrad and Bean at Surveyor 3, the brown color might have been a layer of dust.

To get back on the time line, Mission Control advised the explorers to eliminate two planned stops for samples and photographs. Fine, they agreed. Mitchell chipped fragments off a big rock and Shepard picked up some pieces. Then they moved downhill toward Weird Crater where Houston wanted soil and rock samples.

Meanwhile, flying overhead in Kitty Hawk, Roosa had been busy taking photographs of surface features. Houston was particularly interested in a region of the highlands called Descartes, the probable landing site of Apollo 16.

Moving westward, their visual problems increased. It was well into the second day of the lunar morning, and the Sun was high enough to wash out much of the shadow contrast that showed potholes and craters. Without contrast, craters were hard to see and landmarks began to fade into the general topography.

The Six-Iron Experiment

Shepard told Haise, "This country is so rolling and undulating, Fred, with rises and dips everywhere that you can be going by a fairly good sized crater and not even recognize it."

They reached Weird Crater, or one they were reasonably certain was Weird, and collected rock samples. Houston asked them to make a 25-minute stop at a string of three craters to the west designated on the map as Triplet. There they would join three of the 18-inch core tubes and attempt to get a core sample 54 inches deep by pounding the core tubes into the ground with a mallet. They would also dig a trench 18 inches deep and photograph the strata on the sides of the trench.

When they reached Triplet, they went to work immediately. Unpacking a shovel from the MET, Shepard dug the trench. Mitchell fitted the core tubes together and hammered them into the ground. Both met problems. At a depth of 18 inches, Mitchell found the resistance to further penetration so stiff that he thought he had struck rock. Shepard's trench kept caving in before he could photograph the side walls. The best he could do was to take pictures of the bottom of the trench where there appeared to be a layer of small pebbles that were somewhat glassy in texture.

Near Triplet they found a pile of rocks and bagged some of them. At that point, Mission Control advised it was time "to wrap this one up."

When they returned to the landing site, they checked the alignment of the ALSEP instruments and the central station antenna. They turned on the television camera so that Houston could see their close-out activities. Now and then, one of them skipped in front of the lens, seeming to glide across the field of view. When they had finished these chores, Alan Shepard uncorked his own surprise experiment.

SHEPARD: Houston, you might recognize what I have in my hand as the handle of the contingency sample return, and it just so happens to have a genuine six-iron on the bottom of it. In my left hand, I have a little white pellet that's familiar to millions of Americans.

It was a golf ball.

SHEPARD: I drop it down. Unfortunately, the suit is so stiff I can't do this with two hands, but I'm going to try a little sand trap shot here.
Shepard swung the six-iron one-handed at the ball. He missed.

MITCHELL: Hey, you got more dirt than ball that time.

SHEPARD: I got more dirt than ball. Here we go again.

This time, he connected and the ball traveled about 100 yards.

HAISE: That looked like a slice to me, Al.

Shepard produced another ball and prepared to tee off again, one-handed.

SHEPARD: Here we go. Straight as a die. One more.

He swung and the ball sailed away.

SHEPARD: Ah! Miles and miles and miles!

In the light lunar gravity, the ball traveled 400 yards, Shepard later estimated.[6]

"Very good, Al," said Haise in a 'let's get back to work' voice.

"Al Shepard becomes the first lunar golfer," the NASA Public Affairs commentator announced.

It was 6:38 A.M., Saturday, in Houston. Four hours and 18 minutes of the second EVA had elapsed. Shepard said: "Okay, Houston. Crew of Antares is leaving Fra Mauro base."

The tiny cabin of the LM ascent stage was so crowded with sample bags and gear the astronauts could hardly move. They had brought aboard 43 kilograms (94.6 pounds) of Moon rocks and soil.

Officially, the second EVA had lasted 4 hours and 20 minutes.[7] It brought the total time Shepard had spent outside the LM to 9 hours and 9 minutes, a new record. Mitchell's total EVA time was a few minutes less.

On Apollo 12, total EVA time was 7 hours and 50 minutes, and on Apollo 11, 2 hours and 32 minutes.

At 12:48 P.M., CST, February 6, after 34.2 hours on the Moon, Antares lifted off and made a one-orbit rendezvous with Kitty Hawk, a feat of navigation executed for the first time in lunar orbit. The technique of a single-orbit rendezvous had been worked out on the flight of Gemini 11 by Conrad and Gordon on September 12, 1966. It now paid off on the lunar journey, greatly shortening maneuvering time.

Shepard and Mitchell had no trouble this time docking with Kitty Hawk at 2:35 P.M., 1 hour and 47 minutes after lifting off Fra Mauro station.

During the coast back to Earth, the crewmen began a series of zero gravity manufacturing experiments that were to be enlarged on later Apollo missions and on the subsequent Skylab space station. These experiments anticipated the day when some manufacturing processes might be carried out profitably in a zero gravity environment. Examples were the preparation of pure medicines and vaccines or pure metal alloys, where mixing would not be affected by gravity. One test was designed to determine the effect of eliminating gravitational convection and sedimentation in mixing organic materials. The mixing process was photographed. Another test observed the transfer of heat in liquids and gases in zero gravity. Another was to determine how well metal castings could be made in zero gravity. Some of this work was exhibited in a live color telecast from Kitty Hawk as it flew home.

The industrial tests, programmed by engineers at NASA's Marshall Space Flight Center, Huntsville, Alabama, and the Lewis Research Center, Cleveland, Ohio, were far ahead of their time, perhaps, but they were a practical way of assessing a vision. On the outcome of these and later experiments would hang the feasibility of constructing a large space station in orbit.

The final test, in which the astronauts themselves were the laboratory animals, came after splashdown in the Pacific Ocean on February 9, 1971. Shepard, Mitchell, and Roosa had to undergo the 21-day quarantine invoked by public health scientists to guard against extraterrestrial pathogenic contamination. There had been some doubt that the Fra Mauro formation was as sterile as the maria in Tranquillitatis and Procellarum.

The quarantine that began with lunar liftoff on February 6 was completed on February 27 in the Lunar Receiving Laboratory at Houston. No sign of any organism, living or fossil, was found in the Fra Mauro samples, and the crewmen went home after contributing their negative evidence to the search for life on the Moon.

The Apollo scientific corps now accepted the fact that life had never existed on the Moon until man and his artifacts landed there.

Notes

1. Wood, J. A., et al., "Lunar Highlands Anorthosite and Its Implication," Smithsonian Institution Astrophysical Observatory, Apollo 11 Lunar Science Confernce, 1970.

2. Morison, Samuel Eliot, *The European Discovery of America*, Oxford University Press, New York, 1971.

3. Latham, Gary, Apollo 14 Prelaunch Science Briefing, Jan. 29, 1971.

4. UPI Report from Moscow, *The New York Times*, February 5, 1971.

5. Active Seismic Experiment, Apollo 14 Preliminary Science Report, NASA, 1971.

6. Post-Apollo 14 News Briefing, Washington, D.C., NASA, March 1, 1971.

7. Apollo 14 Preliminary Science Report.

9

Hadley-Apennines

On the Apollo flights, one of the principal experiments in which the astronauts were the subjects was biomedical. The effect of zero gravity on the human organism had been under continuous investigation in the American and Russian space programs since the beginning. Its effects could not be predicted; they had to be observed through experience.

Zero gravity is a misnomer. There is no part of the Solar System, the galaxy, or the universe, for that matter, where gravitational force is zero. Theoretically, there are points where gravitational forces may balance to create a neutral zone, but none where the force is nonexistent.

In orbital flight a vehicle and its occupants are tied to the planet by gravity. On interplanetary flights they are tied to the Sun. And on a flight beyond the Solar System vehicles such as Pioneer 10 and 11 are tied to the gravitational string of the galactic center.

But the term zero G has come into the common language as a synonym for weightlessness, a result of free fall. This is what spaceship crews experience as their vessels literally fall around a planet they are orbiting. The parachutist feels weightlessness for the few moments of free fall before his parachute opens. The weightless condition can be achieved in an airplane when its downward swooping acceleration gives the occupant a fleeting impression of being afloat. This effect was used for the early training of astronauts.

Of all the environmental conditions in space, free fall or zero G is the only one that could not be simulated for any length of time on Earth. The question that has persisted throughout the short history of space exploration is how well the human being fares over long periods of weightlessness, inasmuch as the organism is designed to withstand 1 *G* stress.

By the end of 1970 only a partial answer was in. In 9 years of orbital flight, Americans had flown missions up to 14 days (Gemini 7 in 1965) and Russians up to 18 days (Soyuz 9 in 1970). None of the missions had resulted in any lasting physiological effects, but there were short-term changes that indicated a definite adaptation to the zero *G* condition.

The cardiovascular system underwent temporary deconditioning; there was a loss of muscle mass and, to a lesser extent, bone mass. There was a loss of fluids, which showed up as weight loss immediately after landing. There were losses in red blood cell mass and in plasma volume.

All the American astronauts exhibited a condition termed "negative lower body pressure" during post-landing tests. There was a pooling of blood in the lower trunk and legs when they first returned to 1 *G*. It indicated a loss of cardiac and vascular efficiency and frequently produced symptoms of vertigo. There was a suspicion that the loss in muscle mass extended to the heart muscle.

Within a period of 14 days after landing, however, the Gemini and Apollo astronauts returned to their preflight condition. They quickly regained lost weight, normal fluid balance, and, over a longer period, normal cardiac functioning.

Measurements of body fluids and analyses of excretia returned from flights showed that bone and muscle tissue were lost by excretion during flight. It was as though the body adapted to the lack of gravitational stress by ridding itself of mass it didn't require in a weightless condition.

A medical team under the supervision of Dr. Charles A. Berry, NASA Director of Life Sciences, sought a hypothesis that would explain biochemical mechanisms of the zero *G* adaptation syndrome. Dr. Berry had been the astronauts' physician since Project Mercury. His "bedside manner" during flight was a model of calmness and reassurance. There were minor medical problems, such as dizziness and nausea, during flights, but no serious injury or illness ever developed in space, or on the Moon.

Berry and his medical associates believed that if they could ascertain the biochemical impact of weightlessness, they would be able to control its effects on long space journeys. They might also be able to determine a point at which the adaptation was completed, the point at which the body's biochemistry achieved a balance with the new environment. The big question was: If there was such a point, what was the condition of the astronaut when it was reached? Would he be able to stand when he landed on Mars after an 8-month flight? Or would the continuing physiological effects of prolonged exposure to zero *G* forever keep mankind restricted to brief space flights without some form of artificial gravity— a force that would restore weight. Rotating space stations had been envisioned at the Marshall Space Flight Center, but the consensus of engineers there was that artificial gravity was beyond the state of space technology in the Apollo decade. Also, its development would be too expensive to contemplate realistically in a program already under fire for its high cost.

Although weight loss in flight resulted mainly from fluid loss, it was not clear to what extent it represented depletion of extracellular or intracellular water, and how the depletion related to loss of tissue mass.

On Apollo 14 new measurements were applied to get data on this question. When the results were added to those of earlier flights, the medical team was able to construct a hypothesis of human adaptation to space flight.[1]

There were several anomalies on Apollo 14. Shepard, Mitchell, and Roosa were exposed to the highest radiation doses of any Apollo crew, a total of 1.4 rads* during the 9-day mission. Their translunar trajectory took them into the heart of the Van Allen radiation zone and they flew during a period when low solar activity permitted a high flux of galactic cosmic rays to invade the inner Solar System. The crewmen saw light flashes when their eyes were closed on the average of once every 2 minutes during the outbound and return flights. The flashes, as mentioned earlier, are produced by the impact of cosmic-ray particles on the optic nerve. Several flashes were seen by two crewmen at the same time, showing that high-energy particles not only zipped through the spaceship hull but also through the head of one man and entered the head of his mate. According to the medical report no discernible damage was done by the radiation exposure, equivalent to about two or three medical X-rays.

The second anomaly was that Shepard gained a pound on the flight, the first astronaut to do so. Mitchell lost a pound. Roosa, who remained in lunar orbit, lost 12 pounds. This seemed to illuminate the effect of the Moon's one-sixth *G* in altering the adaptation syndrome. It appeared that the 34 hours and 11 minutes that Shepard and Mitchell spent on the lunar surface greatly reduced the effects of the flight on

---

* Rad is an acronym for "radiation absorbed dose." It is the amount of ionizing radiation delivering 100 ergs of energy per gram of absorbing material. Under long-standing Atomic Energy Commission radiation protection standards, the maximum permissible dose for a population is 5 rads per person over a period of 30 years, or 0.170 rads a year.

them, compared with Roosa, who remained in zero gravity for the entire 217 hours of the mission.

Roosa also exhibited more pronounced cardiovascular deconditioning than either of his teammates. The postflight analysis showed that he lost 9.1 percent of red blood cell mass, compared to losses of 1.7 percent for Shepard, and 4 percent for Mitchell. Roosa also lost 9.7 percent of plasma volume, whereas Shepard gained 1.2 percent and Mitchell 0.1 percent.

Measurements of total body water showed losses of intracellular fluid, but no significant loss of extracellular fluid. Berry concluded that the body water deficit seen immediately postflight must be caused by a decrease in intracellular fluid volume.[2] Shepard lost 2.7 percent, Mitchell 2.6 percent, and Roosa 2.7 percent of intracellular fluid, compared to preflight levels.

From these and earlier data it was theorized that upon initial entry into weightlessness there is a total redistribution of circulating blood volume.[3] There is an increase in the filling of the heart's right chamber and it is interpreted by the body as a need to reduce total fluid volume. Hormone changes occur. Antidiuretic and aldosterone hormones decrease, causing loss of water, sodium, and potassium through the kidneys and a consequent drop in body weight.

The response is a decrease in plasma volume and a reversal of the earlier decrease in aldosterone and antidiuretic hormones. At this point, the body enters a phase of electrolyte and fluid imbalance, in which sodium levels increase whereas potassium levels fall.

Berry and his associates believed that the response to the new fluid and electrolyte condition is an intracellular exchange of potassium and hydrogen ions. Bone density and muscle mass decrease. Cells become hyperacidic. The respiratory system is stimulated to decrease carbon dioxide in the plasma by increasing the breathing rate. The adaptation now reaches its presumed final phase. Potassium is reabsorbed in the renal tubules (of the kidneys). Loss of body weight halts. The body now stabilizes with a new load on the cardiovascular system and a new fluid and electrolyte balance—which is negative to that existing preflight. The analysis concluded: "It is reasonable to presume that the new, stabilized condition is appropriate for long-term existence under weightlessness. The extent to which it is appropriate for work activities on planetary surfaces or for sudden and vigorous return to the unit gravity of Earth are questions remaining to be answered."[4]

Thus, from this hypothesis, it seemed that men could remain in zero gravity indefinitely. Whether they could stand up when they reached Mars after an 8-month flight remained in doubt.

Endeavor and Falcon

The success of Apollo 14 boosted confidence and morale in the far-flung NASA organization and its satellite contractors. Space exploration had become an industry in 10 years. With the government the only customer, however, the industry was subject to wild fluctuations with the national mood. Early in 1971 program cuts required by a budget of $3.3 billion were causing extensive lay-offs in the NASA-industrial work force. The agency's budget had dropped from $3.85 billion in the 1970 fiscal year to $3.3 billion in the fiscal year ending June 30, 1972, a cut of $550 million.[5] It was then to be stabilized at a level of $3.2 billion for the next several years, permitting development of the reusable ground-to-orbit rocket plane called the Shuttle as the Apollo program was phased out.

With descending space funds employment fell rapidly. The total working force employed in NASA programs by government and industry (not counting the military) had peaked at 420,000 in June 1966. By June 1970 it had fallen to 145,000.[6] The cutback contributed heavily to the scientist–engineer recession of 1970–1971, when thousands of highly trained and talented men and women who had created the means of space travel found themselves out on the bricks, seeking more mundane employment. Technical teams of great brilliance and imagination were dispersed, the training of graduate students in science in the universities, once handsomely funded by NASA, was cut and eventually discarded. NASA funding of space science centers and laboratories stopped, and research grants were sharply curtailed. At the height of the exploration of the Moon the national retreat from the Moon was being programmed. By 1971 the Apollo transportation system had become obsolete and the end of manned exploration of the Moon, at least in the twentieth century, was only a year away.

Six months after the launch of Apollo 14, a new crew and a new spacecraft were ready for launching at the Kennedy Space Center on July 26, 1971. Apollo 15 was destined to make the most ambitious voyage yet.

The Apollo spacecraft had been awarded the call sign Endeavour, after the ship in which Captain James Cook, another seeker of the Northwest Passage, circumnavigated the Earth in 1768–1771. The lunar module was named Falcon. The crew of Apollo 15 was one of the most academically advanced of the Apollo crews. All had master's degrees in aeronautics and in astronautics. All three were service academy graduates and Air Force officers.

The commander was Colonel David R. Scott, 39, the son of one retired Air Force brigadier general, Tom W. Scott, and the son-in-law of another, Isaac W. Ott. A tall, rangy Texan, Scott and his wife, Ann, had two children, Tracy, 10, and Douglas, 7.

Scott was a graduate of the U.S. Military Academy. He had received a Master of Science degree in Aeronautics and Astronautics from the Massachusetts Institute of Technology. He was regarded as a top rank pilot. He flew the Gemini 8 mission with Neil Armstrong on March 16, 1966. As related earlier, the two made the first rendezvous and docking with another vehicle (an Agena rocket). Minutes after this accomplishment, however, an attitude control thruster on the Gemini capsule began firing out of control and put the joined vehicles into a spin. The astronauts were forced to undock in order to bring the Gemini under control and then make an emergency landing.

Scott's second space mission was more successful. He was command module pilot of Apollo 9, an Earth-orbit mission in which Air Force Colonel James A. McDivitt, the mission commander, and Russell Schweickart, an Air Force pilot turned physicist, tested successfully the lunar module in Earth orbit.

Scott's partner on the Moon was Lieutenant Colonel James B. Irwin, 41, a U.S. Naval Academy graduate who had joined the Air Force. He had a Master's Degree in Aeronautical Engineering and Instrumentation from the University of Michigan. Born in Pittsburgh, Irwin called Colorado Springs his home. He and his wife, Mary Ellen, had four children: Joy, 11; Jill, 10; James, 8; and Jan, 6. Irwin was lunar module pilot.

Neither Irwin nor the Command Module pilot, Major Alfred M. Worden, 39, had flown in space before. Like Scott, Worden was a graduate of the U.S. Military Academy. Also, like Irwin, he had done graduate work at the University of Michigan where he received a Master's Degree in Aeronautical Engineering and Instrumentation and in Astronautical Engineering and Instrumentation. Worden had served as an instructor at the Aerospace Research Pilots School, from which a number of astronauts had been drawn. He and his former wife, née Pamela Ellen Vander Beek, were parents of two children, Merrill E., 13; and Alison P., 11.

The Rover

The program of their 12-day mission increased surface stay time from 34 to 67 hours and EVA time from 9 to 20 hours, with three excursions instead of two on the Moon and a fourth EVA by Worden on the return journey to retrieve photographic film on the service module.

Apollo 15 carried the first, self-propelled land vehicle to the Moon, the four-wheeled, battery-powered Lunar Roving Vehicle (LRV). With a cruise radius of about 20 miles, this extraterrestrial jeep greatly extended the range of exploration from the lunar module—hitherto limited to about 1 mile. Because of life-support oxygen limitation, Scott and Irwin were restricted to a radius of 6 miles, the maximum distance they could walk back to the LM if the jeep broke down.

The LRV, or Rover, was 10 feet 2 inches long and 44.8 inches high, and had a 6-foot tread width, and a 7.5 foot wheelbase. The wheels, 32 inches in diameter, were constructed of a woven mesh made of zinc-coated piano wire with chevron-shaped treads of titanium. Each wheel was powered individually by a 0.25-horsepower motor, giving the Rover a total of 1 horsepower and a top speed of 8 miles per hour on a smooth surface. Power was supplied by two 36-volt batteries. Weighing 460 pounds (Earth), the Rover had a Moon weight of only 76 pounds and one man could lift it. It could haul 1,080 pounds (Earth) including the two astronauts and their gear weighing a total of 800 pounds: 100 pounds of communications equipment, featuring a mobile television broadcast unit and antenna, 120 pounds of scientific equipment, and 60 pounds of lunar samples.

The Rover television camera could be operated by radio control from Houston as well as on the vehicle, providing the first pictures of a lunar traverse. A second camera was to be mounted at the landing site.

The vehicle traveled to the Moon in a bay at the base of the LM with its wheels folded against the chassis. It was pulled out, wheels first, like a Murphy bed after lunar landing.

The Science Mission

Scientific experimentation on Apollo 15 was more ambitious than on the earlier flights. It began as soon as the crew achieved Earth parking orbit. During that time the astronauts were to photograph the Earth's surface in ultraviolet light and to continue the ultraviolet photography at various distances throughout the voyage to the Moon. One of the photographic objectives was to establish the utility of photographing the surface from a space station in the ultraviolet and infrared regions as well as the visible light portion of the spectrum.

In lunar orbit a series of experiments with gamma-ray and X-ray detectors was planned to survey the broad chemical provinces of the surface. Extensive photo mapping was scheduled, most of which would be done by Worden while Scott and Irwin were on the ground.

The service module segment of the Endeavour carried a 31-inch subsatellite, shaped like a rural mailbox, to be pushed out of the vehicle and into its own orbit by springs. It would remain in lunar orbit long after Endeavour had departed for Earth. The subsatellite carried an S-band transponder so that its orbit could be tracked from Earth, a magnetometer, and particle detectors to measure plasmas and particles in the Moon's low altitude environment.

Orbital tracking of the subsatellite, like tracking of the Apollo space-craft itself, provided data on the gravitational "anomalies"—the "highs" and "lows" of the gravity field corresponding to the arrangements of mass in the Moon. In this way tracking of the Lunar Orbiter cameras in 1966–1967 revealed the existence of mass concentrations (mascons) under the circular maria. At the time of Apollo 15 it was generally held that the mascons were congealed pools of lava from the melting epi-sodes.

Enroute back to Earth, the crew would be kept busy by continuing the industrial experiments in zero gravity begun on Apollo 14. Of par-ticular interest here was the effect of weightlessness in making metal alloys. Samples of indium and bismuth were to be melted in a small resistance heater, allowed to cool, and then examined.

On the ground, Scott and Irwin were to investigate the first moun-tains men had studied on the Moon and explore the great, serpentine Hadley Rille to try to determine what it represented. From afar it resembled a river gorge, but, in the absence of water, the only fluid it might have carried would have been molten lava. Was it a collapsed lava tube? Or simply a crack in the crust?

Lava Lakes

During the late spring and early summer of 1971 new seismic infor-mation from Apollo 14 had been digested. Now there were two stations listening for sound waves traveling through the Moon, one at Fra Mauro and an earlier one in the Ocean of Storms, 108 miles apart.

The additional seismic information showed that Moonquakes were not confined to shallow depths but were generated 30 to 400 miles below the surface. The maria appeared to be surrounded by a heteroge-nous zone that scattered the seismic waves, but below it wave transmis-sion was so efficient that Latham and his colleagues were convinced they were seeing impacts of meteorites from all parts of the Moon—and not just within a few hundred kilometers of the station as they had thought earlier. About 20 percent of the impacts were from rocks the size of grapefruits striking the far side of the Moon, they concluded.

It began to appear that the Moon had a crust, but its nature and extent could not be determined at that time. The depth of the basaltic lava flows was still undetermined. They could be only "skin deep" or extend to a depth of 30 to 60 miles.

Moonquakes were not being detected regularly. They seemed to occur in 11 different zones. The "hottest" zone, accounting for 80 percent of quakes, was about 400 miles distant and at considerable depth. When the Apollo 12 seismometer was operating alone, the hot spot appeared to be in the vicinity of Fra Mauro, but after the Apollo 14 seismometer joined the "network," the hot spot's apparent location moved farther south and west.

It was clearer now that the quakes recurred at monthly intervals, like clockwork. Two events a month were recorded 5 days before the Moon reached perigee in its orbit and one event came 3 days before perigee. The precision was astonishing. "You can nearly set your watch by it," said Gary Latham.[7]

Although these tidal Moonquakes differed from most earthquakes, they did resemble terrestrial volcanic activity in one respect. They were analogous to earthquakes observed in Hawaii—a type of quake produced by a reservoir of magma 50 or 60 miles down.

The similarity suggested to Latham that the Moonquakes were also produced by a magma pool below the surface. The seismic team supposed that the effect of the increased attraction of the Earth at perigee was to pump molten fluid into cracks between the rocks and push them apart. The fracturing of the rock was then observed as a Moonquake by the seismometers. It was estimated that a magma reservoir only a few kilometers deep could produce this effect.

The possibility that the lunar quake zones were also sources of the luminous hazes seen from Earth occurred to investigators. In his report, Kozyrev had said that the central peak of the Crater Alphonsus in the highlands east of Mare Nubium became washed out and acquired a reddish hue as he observed it November 3, 1958. Hazes had been seen in the bottom of Alphonsus earlier—also in the Craters Eratosthenes and Plato. Were these gases? Kozyrev thought so.[8] He proposed that they represented an emission of volcanic ash or dust, followed by several hundred thousand cubic meters of gas for a period of several hours.

If the Moon had once been molten, it was reasonable to believe that some molten material remained even if most of the body had cooled. One of principal sites where supposed gas emissions had been observed by watchers on Earth was the Crater Aristarchus in Oceanus Procellarum. Among Worden's assignments in lunar orbit was to search Aristarchus with gamma-ray and alpha-particle detectors for signs of radioactivity and gases.

Gases of probable volcanic origin had been detected by the cold cathode gauge set up by Apollo 14. The instrument was designed to measure the tenuous lunar atmosphere at the surface. An earlier gauge set up by Apollo 12 had failed.

Although there was no "sensible" atmosphere on the Moon, there was, technically, an atmosphere that amounted to about 1 million particles per cubic inch. The Apollo 14 gauge sensed concentrations of 100 million particles per cubic inch—vertitable gas clouds compared with the usual concentrations. Once the experimenters decided that the gases did not come from the LM, which had emitted gases for a while, they tried to figure out the source of concentrations the gauge was reporting. This was not possible with only one gauge on the Moon.

The gauge did not indicate the chemical composition of the gases. However, Kozyrev's spectrograms of emissions from Alphonsus showed carbon-2 and carbon-3 molecules, indicating the breakdown of some carbon compound.

In the Marsh of Decay

Curiously enough, the Marsh of Decay (Palus Putredinus) suggested a region where gaseous hazes might have been seen. The landfall of Apollo 15 was a basaltic embayment on the eastern ridge of the Imbrium Basin, 640 miles northeast of the Fra Mauro formation. The landing target was at 3 degrees, 39 minutes, 10 seconds east longitude and 26 degrees, 6 minutes, 4 seconds north latitude. It lay in a roughly triangular patch of mare material between the Apennine Mountain front and the sinuous canyon-like Hadley Rille, 350 meters (1,137 feet) deep and 923 meters (3,030 feet) across.

Towering 6,000 to 15,000 feet above the landing site, the Apennines formed the main, southeastern rim of Imbrium. The center of the basin lay 390 miles to the northwest.

Pushed up by impact, these mountains are a chain of large massifs. The huge blocks were lifted out of the Moon and piled up at the basin rim when the planetesimal struck 3.9 to 4 billion years ago.

Two major massifs towered over the Putredinus plain. They had been named Mt. Hadley and Mt. Hadley Delta. Looming up northeast of the landing site, Mt. Hadley is 4.5 kilometers (14,625 feet) high and Hadley Delta, south of the site, is 3.5 kilometers (11,375 feet) in altitude.

The landing site put the explorers near enough to the mountain front and the rille to explore both areas, using the Rover for transportation.

Three EVAs were planned to explore the front, the rille, and the plain itself. The ALSEP this time included the complicated heat flow experiment, which had been lost on Apollo 13.

The Rounded Mountains

Apollo 15 was launched on time at 9:34 A.M., Eastern Daylight Time, July 26, 1971, from the Kennedy Space Center and entered lunar orbit at 3:05 P.M., Houston time (Central Daylight Time), July 29. Once the spacecraft was off the pad, control was assumed by the Manned Spacecraft Center, Houston, and events were clocked by Central Daylight and by Greenwich Mean Time.

Scott and Irwin brought the lunar module Falcon down on a cratered slope between the mountain front and the rille at 5:16 P.M., CDT, July 30. Their first task was to depressurize the cabin and open the top hatch so that Scott could stand up and take a good, long look at what was out there. This was a stand-up EVA, or SEVA, and it lasted 33 minutes, during which time Scott described the scene and photographed it with the 500-millimeter telephoto lens of the Hasselblad camera.

He talked at length to the capsule communicator at Houston, Joseph P. Allen. A physicist, Allen was one of a group of scientists accepted for astronaut training in 1967, when it was expected there would be at least nine lunar landing missions instead of seven. This was the closest he was to get to exploring the Moon.

While Allen monitored Scott and Irwin on the surface, another scientist-astronaut, Karl G. Henize, an astronomer, served as capsule communicator for Worden, who was carrying out the orbital experiments aboard the Endeavour.

"All the features around here are very smooth," Scott said. "The tops of the mountains are rounded off. There are no sharp, jagged peaks or no large boulders apparent anywhere. The whole surface area appears to be smooth." It sounded as though he was looking at the Sahara Desert.

The boulders, as it turned out, were there—in large numbers. But they were not visible in the first, panoramic view of the lunarscape. The scene was as rolling and dune-like as a desert.

After the visual survey, the Moon crew repressurized the cabin and settled down for a supper of freeze-dried and specially processed soups, meats, and vegetables that, when properly hydrated, sometimes tasted pretty good. Then they slung their hammocks and crawled in. As Irwin related later, they never conversed during these rest periods on the Moon. Each was preoccupied with his own thoughts and each was unwilling to disturb the other.[9]

Early in the morning of July 31, Mission Control detected a drop in the cabin oxygen pressure. The flight director on duty, Peter Frank, awakened the crew at 3:59 A.M., an hour ahead of the scheduled morning call, and advised the astronauts of a possible cabin leak. After a search, the crewmen discovered that the urine dump valve was open. When it was closed, the oxygen leak stopped.

"Okay you guys," said Mission Control cheerfully, "You have 22 minutes left if you want to go back to sleep again."

They didn't see any point to that and began to prepare for the first big day at Hadley-Apennines—a day devoted to unloading the Rover, exploring the mountain front near Mt. Hadley Delta, taking a look at the rille, and then, after returning to base, setting up the ALSEP instruments.

Shortly after 8 A.M., the crew opened the MESA (modularized equipment stowage assembly) and deployed the television camera. Houston was delighted with the picture. Setting foot on the surface, Scott made an impromptu speech: "Okay, Houston. As I stand out here, among the wonders of the unknown at Hadley, I sort of realize there's a fundamental truth to our nature. Man must explore. And this is exploration at its greatest."

In the next breath, Scott said the LM was tilted. It stood on a 10-degree slope, with the left rear foot pad 2 feet lower than the right and the left front pad low, too. The bell of the descent engine rested on the rim of a shallow crater.

On this mission, Scott and Irwin wore updated Moon suits. The neck and waist had been convoluted to make the suit more flexible. They could turn, nod their heads, twist, and bend forward more easily than their predecessors. Bending his head back a little and looking up in the black sky, Scott said he could see "a rather interesting sight. I can look straight up and see our good Earth up there."

Unloading the Rover was harder than it had been in practice. There were numerous "take it easy" and "atta boy" and "easy now" phrases floating between the Earth and Moon during that exercise. Finally, Irwin spoke the magic words: "Got it!"

The Moon car appeared on television. It looked like a topless dune buggy with wire wheels.

"Pretty sporty there, Jim," commented Joe Allen in Houston.

"Man, this thing is nice and light," said Irwin.

At the outset the front wheel steering failed to operate. Every effort to engage it failed. The vehicle was also steerable by the rear wheels, so the explorers settled for that. Scott got into the driver's seat and drove the vehicle slowly away from the television camera and out of sight.

In final assembly of the Apollo 14 rocket and spacecraft six months later, an overhead crane lifts the launch escape system (LES) in place atop the Apollo command module. With the LES, the Saturn 5–Apollo "stack" is 363 feet tall. The LES consists of three powerful solid-propellant rockets designed to boost the command module away from the stack and propel it high over the Atlantic Ocean so that it can descend on its parachutes in the event of an emergency on the launchpad. This photograph was made November 8, 1970. Apollo 14 was launched January 31, 1971.

ssembly of the Saturn 5 Moon rocket to launch .pollo 14 to the Fra Mauro formation on the Moon ʝes forward in the giant Vehicle Assembly Building VAB) at the Kennedy Space Center, Florida, with ɪe mating of the Instrument Unit (IU) with the ɔcket's third stage (S4B). This photograph was made ʧay 14, 1970.

The first clue to the basaltic nature of the lunar maria surface was reported by a little gold box aboard Surveyor 5. The box contained a chemical analyzer devised by Anthony Turkevich of the University of Chicago and James H. Patterson (right) of Argonne National Laboratory. Gold plating controlled the temperature inside the instrument, which was lowered automatically to the lunar ground by a light cable, such as the one Professor Turkevich is holding.

The crew of Apollo 11, Neil A. Armstrong, Michael Collins, and Edwin E. Aldrin, Jr., interrupt their training to watch as their space vehicle is placed in position on launchpad 39-A, May 20, 1969, in preparation for launch July 16, 1969. The vehicle has just been moved 3.5 miles to the pad aboard a 3,000-ton, diesel-powered tractor from the Vehicle Assembly Building where rocket and spacecraft were assembled and checked out.

Small, shiny, spherical particles appeared in this stereo camera view of a clump of lunar soil returned from Tranquility Base by Armstrong and Aldrin. The 35-millimeter camera was mounted on a walking stick. Astronauts operated it by holding the base of the stick against the object to be photographed and pulling a trigger at the top of the stick. Rigidity of the Moon suits prevented the explorers from bending over or kneeling to take close-up pictures.

This Apollo 11 close-up view of a 2½-inch rock excited geologists because the tiny pieces around it appeared to indicate erosion. What were the erosive agents on the windless, waterless Moon? The clue is here. On the surface, small pits are visible. Their raised rims are typical of high-velocity, micro-meteoroid impacts.

Engineers and technicians in the Firing Room at Kennedy Space Center monitor their consoles during the Apollo 11 countdown demonstration test. The test preceded the actual countdown for the launch.

Apollo 11 lunar module pilot Aldrin sets up the first seismometer on the Moon. The instrument, designed to detect Moonquakes and the impacts of meteorites, was so sensitive it recorded the astronauts' footfalls and radioed the sounds to Houston.

Aldrin uses a core sampler to obtain a lunar soil sample at Tranquility Base. The instrument is a tube which he forces into the ground by its handle. The dirt in the tube was brought back to the Lunar Receiving Laboratory at Houston for preliminary analysis and some portions were later distributed to investigators in the United States and Europe for chemical analysis. In the background is an aluminum-foil pennant designed to collect particles in the solar wind. The photographer is Neil Armstrong.

Armstrong is photographed by Aldrin beside the seismometer experiment. Solar panels designed to provide power during the long lunar day can be seen on each side of the instrument. In the background is the lunar module, Eagle.

Apollo 12 lunar module pilot Alan Bean begins to remove the Apollo lunar surface experiment package (ALSEP) from the lunar module Intrepid. Resembling an umbrella inside out is the S-band antenna pointing at the Earth.

Bean inspects the TV camera on Surveyor 3. The reconnaissance vehicle had been standing on a crater slope in Oceanus Procellarum since it landed two and one-half years earlier, on April 20, 1967. Bean and Apollo 12 commander Charles Conrad, Jr. found the vehicle covered by a brown film of dust. Beyond the crater rim is the lunar module, Intrepid.

This is how Surveyor 3 looked to the Apollo 12 astronauts as they hiked toward it in November 1969. Conrad and Bean piloted their lunar module, Intrepid, to a pinpoint landing within walking distance of the historic reconnaissance vehicle. The awning-like structures at the top of the vehicle are solar panels that powered the TV camera and other instruments.

Bean, taking the picture, is mirrored in the sun visor of Conrad's helmet in this close-up photo of man at work on the Moon. Conrad is holding a portion of a core sample tube.

Portrait in black and white of Apollo 12 lunar module, Intrepid. It stands to this day as a monument to the voyages of Apollo on the stark lunar plain of Oceanus Procellarum. Beside it is the S-band antenna designed to relay television and voice communications to Earth.

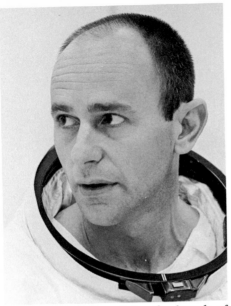

Alan Bean "suiting up" for the launch of Apollo 12.

Apollo 12 command module pilot Richard F. Gordon, Jr. gets ready for the second lunar landing mission.

Prelaunch crew photo of Apollo 12 astronauts at Kennedy Space Center's mission simulator. Foreground to rear: Conrad, Gordon, and Bean.

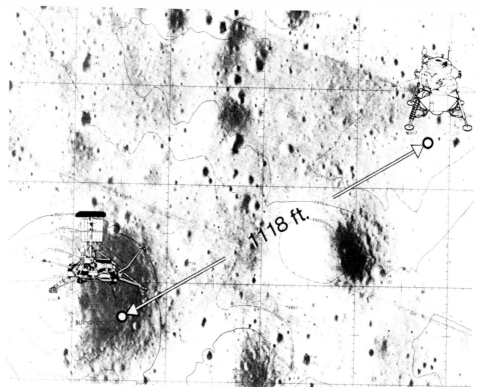

Prime landing site of Apollo 12 is depicted in sketch superimposed on photos made by Lunar Orbiters 1 and 3 three years earlier. The crew was to land within 1,118 feet of Surveyor 3. Intrepid actually landed within 600 feet of Surveyor.

Two of the three Apollo 13 crewmen, Fred W. Haise, Jr., left, lunar module pilot, and James A. Lovell, Jr., commander, relax at breakfast a few hours before launch.

Apollo 13 commander Jim Lovell.

Apollo 13 command module pilot John L. Swigert, Jr.

Project Apollo ends December 19, 1972 with the safe splashdown of Apollo 17 astronauts Eugene A. Cernan, Ronald E. Evans, and Harrison H. Schmitt in the Pacific Ocean southeast of Samoa.

Donald K. Slayton, Director of Flight Crew Operations (seated), and Thomas P. Stafford, chief of the astronaut office, monitor the Apollo 14 countdown demonstration test in firing room 2 at Launch Control, Kennedy Space Center.

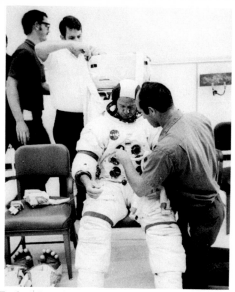

Technicians help astronaut Alan B. Shepard, Jr., Apollo 14 commander, suit up.

Apollo 14 crewmen Edgar D. Mitchell, Alan B. Shepard, Jr., and Stuart A. Roosa (right) pose for photographers as their Saturn 5–Apollo vehicle is moved to the launchpad aboard a giant tractor.

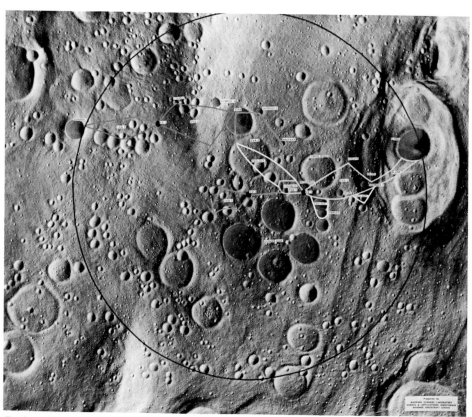

Prelaunch view of the Apollo 13 landing site, the hilly Fra Mauro region of the Moon, shows planned, extra-vehicular activity traverses. The mission never landed.

Dune-like, lunar mountains photographed by the crew of Apollo 15 at the Hadley-Apennines landing site.

One of the most striking photographs of the Apollo voyages is this portrait of James B. Irwin standing beside the lunar rover on the plain before Mount Hadley. The photographer is Apollo 15 commander David R. Scott.

Irwin loads the rover as he and Scott prepare for the first traverse of the Apollo 15 mission.

Scott stands on the slopes of Mount Hadley Delta.

Motoring on the Moon in the battery-powered LRV (lunar roving vehicle), Scott drives the land-going space car first used on Apollo 15.

Charles M. Duke, Jr., Apollo 16 lunar module pilot, was the tenth man to land on the Moon.

The lunar roving vehicle is folded for stowage aboard the Apollo 15 lunar module at the Vehicle Assembly Building, Kennedy Space Center.

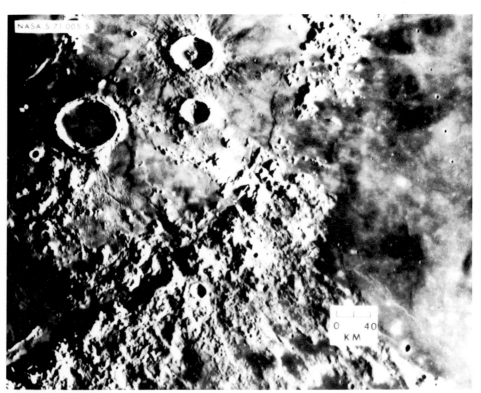

A Lunar Orbiter view of the Apollo 15 landing site in the lunar Apennines. The big crater at the upper left is Archimedes.

HADLEY NORTH

LRV TRAVERSES

- ⬤ 300-350 m
  (ABOUT CONE CRATER DIA)
- ⚬ 100-125 m
  (ABOUT NORTH TRIPLET DIA)
- ∘ 50 m
  (ABOUT FLANK CRATER DIA)

0 1 2 3

KM

Map of the traverses planned for the Apollo 15 mission.

Apollo 15 commander Scott, the United States flag, and the lunar module, photographed by Irwin. The big sandpile in the background is Mount Hadley Delta, 5 kilometers away. It rises about 3.5 kilometers above the plain.

A view of the Apollo 16 landing site in the lunar highlands photographed from the Apollo 14 command module.

Astronaut Thomas K. Mattingly II. Bumped from the ill-fated Apollo 13 mission because of susceptibility to rubella, he was reassigned as command module pilot of the successful Apollo 16 flight.

Apollo 16 commander John W. Young before the flight.

Apollo 16 commander Young at work on the Moon.

Astronaut Harrison H. Schmitt, the only geologist to get to the Moon in Project Apollo, chips away at a boulder at the base of the North Massif during the Apollo 17 mission. At left is the lunar rover and its S-band antenna.

The scene at Taurus-Littrow, a valley in the highlands. Schmitt is shown working at the rover in this photo by Apollo 17 commander Eugene A. Cernan. The famous orange soil which excited geologists is all around the rover.

View of the front side of the Moon with Apollo landing sites marked.

Mission	Latitude	Longitude
Apollo 11	23.5° E	0.6° N
Apollo 12	3.03° S	23.4° W
Apollo 14	3.66° S	17.48° W
Apollo 15	26.12° N	3.65° E
Apollo 16	15.5° E	9.0° S
Apollo 17	20° N	30° E

Lunar exploration, circa 1972. Seated in the lunar rover is Apollo 17 command module pilot Ronald A. Evans. Standing (left) is Schmitt, the lunar module pilot, and Cernan, mission commander. In the background is the Saturn 5–Apollo vehicle system ready for the lunar journey.

"We're moving," he called redundantly.

"Outstanding!" enthused Joe Allen.

By the Rear Wheels, March!

The phrase "outstanding" was a familiar one in the program. Often, it expressed a faintly sardonic response when some Rube Goldberg device or scheme actually worked. But there was no doubt that the image of this funny looking vehicle rolling smoothly along on the Moon, with the blimplike figures sitting in it, was one of the more outstanding sights of the century.

Scott parked the car, and he and Irwin erected its high-gain antenna and the television camera that looked forward and that could be operated and pointed by remote control from Houston.

The camera gave Mission Control closer surveillance of the crew on traverse than was possible by only voice transmissions. In the control room, a wag had put a duplicate of the Rover television remote control panel on the desk of Chris Kraft, then deputy director of the Center, with a card on which was inscribed "Cecil B. DeKraft." It was the only allusion to the marionette aspect of lunar exploration under continuous direction from Earth. The days when explorers vanished into the wilds and were on their own for months or years were gone.

When the camera was turned on, Joe Allen was impressed by the scene, which he described, this time, as "breathtaking."

"Good, Joe," said Scott. "It can't be half as breathtaking as the real thing, I'll tell you. Wish we had time to just stand here and look."

After a considerable amount of difficulty buckling their seat belts over their bulky Moon suits, the explorers rolled off to the southwest, Scott at the controls, steering with the rear wheels.

The ride was a bouncy one. They moved along at 5½ miles an hour, slowing only a little as they rolled up a hill. Joe Allen asked them to head southwest, toward St. George Crater and Mt. Hadley Delta. Almost immediately, Irwin noticed that the cleated wheels did not sink into the regolith as deeply as their boots did. The wheel tread design was first rate.

"Man," said Scott, making a sharp turn to avoid a hole, "I'm going to have to keep my eye on the road. It's really rolling hills, Joe, just like (Apollo) 14. Up and down we go."

The rear wheel steering worked fine. Scott guided the Rover with a T-shaped hand controller, which governed speed, direction, and braking. The Rover had its own navigation system. It indicated heading, the direction and distance between it and the LM at any point, and traverse

APOLLO **15** TRAVERSES

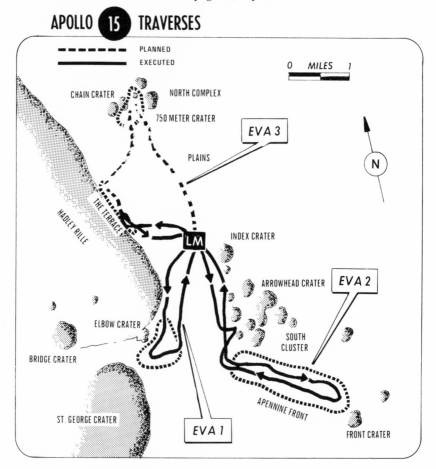

mileage and speed. All the indications were in the metric system, toward which NASA was moving, although the nautical mile still persisted as the unit of distance and velocity in space.

The Rover's navigation system had three components: a directional
gyroscope aligned with the Sun, odometers that recorded distance by
magnetic pulses from each wheel, and a small, solid-state computer that
determined heading, bearing, range, distance traveled, and speed. All
readings were displayed on the control console. At the beginning of each
traverse, the system was reset by pressing a reset button that returned all
the displays to zero.

Built by the Boeing Company under the supervision of the Marshall
Space Flight Center, the Rover essentially was a wheeled spacecraft. It
was developed in 16½ months after Boeing received the go-ahead from
NASA in May 1969. The three vehicles built for Apollo 15, 16, and
17 plus the training vehicles and spares cost $21 million. Although this

seemed to be top dollar for a car, it should be considered that the Rover represented a new spacecraft technology—one that would outlast Apollo. For if the Rover could operate on the Moon, it could probably operate on any planetary surface where man could land.

The only trouble was when something went wrong with this multi-million dollar vehicle, it was difficult to reach the dealer. But as the mission progressed, even this problem was solved, in a most mysterious way.

Moving at 6 miles per hour to the southwest, after leaving the immediate area of the landing site, Scott and Irwin approached the rim of Hadley Rille. Scott reported that the rear wheel steering was quite responsive. Further, there was no accumulation of dirt in the wire wheels as feared.

"Just like the owner's manual, Dave," said Allen.

"This is really a rocking, rolling ride," Scott said. "Never been on a ride like this before. I'm glad they've got this great suspension system on this thing. Hey, you can see the rille!"

Scott continued: "We're looking down at it. Down and across the rille we see craters on the far side. We're getting into the rocky stuff, angular, irregular surface."

Downward, they saw huge, angular blocks on the slope all the way to the U-shaped bottom. One big rock standing on the rim appeared to Scott to be 10 meters across.

Moving southward along the rim, they reached a landmark on the map called Elbow Crater, which seemed to be ancient and subdued. They were nearing their objective—the vicinity of St. George Crater where the splash-out might consist of primeval rocks. Scott continued following the rim of the rille, which wound toward the southwest. At one point the slope of the rille seemed so gentle that Scott commented, "It almost looks like we could drive on this side."

"Stand by on that," said Allen sharply.

"We could drive back out," Scott went on.

The proposal to drive into the rille never got beyond this tentative suggestion. It was not part of the plan.

"Oh, this is really a sporty driving course," said Irwin. "Man oh man, what a Grand Prix this is!"

"This is traveling," Scott agreed. "It's a great sport. I tell you—the sand pile [practice range] was never like this. I wish we could just sit down and play with the rocks a while. Look at these things. They are shiny and sparkly. Look at these babies here. Gosh."

In the splash-out from St. George Crater, they took a sample from a boulder. When they turned it over, they found the underside was covered with glass, some of it having frothed and bubbled. They took a double core (about 36 inches) of the soil. With this, they had gathered samples of all three geological formations in the region—the basalt of Putridenus, the edge of the rille, and, now, the mountain front.

St. George Crater was their farthest south on this traverse. They turned the Rover around and headed back toward Falcon, which was not visible any longer. It was easy to get lost in this hilly and cratered region, but the navigation system was supposed to be accurate and should get them back promptly. However, Allen suggested they try to pick up their outbound tracks and follow them back.

"Oh, the Hansel and Gretel trick," said Scott. "Okay. Say, we see a neat place to go down into the rille."

Again, there was no encouragement for this variation of the plan at Houston, and the subject was not pursued. As they rolled northeast, they topped a rise and spotted Falcon leaning toward them in the distance, about 3 miles away.

Mission Control notified the explorers that their rate of oxygen consumption had been higher than planned. The EVA would have to be shortened about 30 minutes from the planned 7 hours.

Back at Hadley Base, after a drive of 6½ miles, Scott and Irwin parked the Rover and, under the eye of the television camera, set up the magnetometer, the seismometer package, the aluminum foil solar wind collector, the solar wind analyzer, a new and larger laser mirror, which now made a network of three U.S. reflectors on the Moon,* a dust detector, and the central power and broadcasting station for these instruments.

The heat flow experiment presented unexpected difficulties. The experimenters had predicted the two 10-foot holes into which the probes were to be inserted could be drilled in 15 minutes each with the powered jackhammer-style lunar surface drill. This turned out to be wishful thinking. The drilling of the bore holes became so difficult after 3 feet of penetration that the astronauts had to leave the project unfinished until the next day in order to stay within their oxygen limits.

Thus, the first EVA was ended after 6 hours, 34 minutes, and 14 seconds, about 26 minutes early. As the explorers climbed back into Falcon's cabin, Joe Allen passed up a cheery word:

"Real fine day's work up there, guys. Why don't you take the rest of the day off?"

"Okay," said Scott. "Thanks, Boss."

---

* Two previous laser retroreflectors (mirrors) had been set up in Mare Tranquillitatis and Oceanus Procellarum by the crews of Apollos 11 and 14.

The Long Distance Repair

August 1, 1971 was a Sunday at home, but on Hadley Base it was just another working day. The second EVA began at 6:48 A.M., and lasted 7 hours and 12 minutes. Scott and Irwin drove in a southeasterly direction and part way up the slope of Hadley Delta. The Rover climbed easily.

But before they took off for the second traverse, the Rover somehow got repaired. No one is sure how, to this day. All the record shows is that Joe Allen called up and said: "Okay, Dave. Once again we want you to exercise the forward steering procedure here. The forward steering switch should be at buss Charlie and the forward steering circuit breaker should be closed."

Scott followed this procedure. The forward steering suddenly came to life. Astonished, he reported: "You know what I bet you did last night, Joe? You let some of those Marshall guys come up here and fix it, didn't you?"

"They've been working," said Allen, cryptically. "That's for sure."

"It works, Dave," said Irwin, amazed. "It's working, my friend."

"Beautiful," said Scott.

"Lots of smiles on that one, Dave," said Allen. "We might as well use it today."

Scott suggested that maybe the Boeing Company had a secret booster to fly mechanics to the Moon to service its Moon car. Whatever had happened, the steering was much better, he said.

They rolled away in the brightening lunar morning toward the southeast, passing a crater designated as Earthlight on their map. To the east of it lay a little valley they had not seen before. They passed Crescent and Dune Craters and presently they found themselves in a field of rock debris. The fragments were 6 to 12 inches long for the most part, but some ranged up to 2 feet. Scott had to maneuver in a series of sharp turns to avoid the bigger rocks.

The Rover performed beautifully, he reported, moving up a 7-percent grade "like it knows what it's doing." Irwin photographed the scene with a 16-millimeter movie camera. The Rover television was shut off while the car was in motion and turned on during stops.

Turning west, Scott drove around the rim of Dune Crater. Big blocks, some 3 meters across, rose on the southern slope. Beyond the crater loomed the smooth mass of Mt. Hadley Delta. From a distance, it did not look 11,375 feet high, but as the explorers approached its flank, its mass became awesome.

Nothing like this mountain existed on Earth, at least in the Cenozoic

Era, and possibly not since the Precambrian. Although huge planetesimals must have hit the Earth when it was young, the dynamics of Earth land-forming processes had long ago erased the effects. Even a basin on the order of Imbrium would have vanished or would have become a sea. Earth's mountains in the Cenozoic were believed to be created by the compression of continental plates butting up against each other. They were folds, similar to the folds of a rug when the edge is pushed inward.

The mountains of the Moon had been created by compression, too, not of drifting continents; an alien body crashed into the surface and heaved upward some of the material it displaced. They were much more ancient and more gentle than the mountains of Earth.

The base of Hadley Delta was pocked with craters. The explorers paused for a drink of water from a quart bag inside their helmets with a tube, which could be reached by turning the head. Dead ahead was another landmark, Spur Crater, where they hoped to pick up samples of ancient rock, possibly Imbrium ejecta.

After a short break, they drove up to the front. The Sun had risen to 45 degrees above the horizon and the glare made it harder to see than before, especially to discern differences in rock types.

"Boy, that's a big mountain when you're down here looking up," Scott observed.

Ahead of them was a line of four craters going up the slope. Each was about four meters across. Below lay a big boulder. It was tempting to imagine that the boulder had made the craters, bouncing out of one and then impacting below it as it caromed down the mountain.

The uphill drive was easy and the slope seemed to be no more than 3 to 5 degrees. As their elevation increased, they got a better view of Mt. Hadley to the north. Irwin remarked that as it became illuminated in sunlight, Mt. Hadley showed distinct, linear markings, as regular as stripes, dipping toward the northwest at an angle of 45 degrees. On Earth, these would be interpreted as sedimentary deposits but there was no known analogous process on the Moon. The lineaments became more distinct as they moved upward on Hadley Delta.

The Banded Mountains

"By golly, Joe. This Rover is remarkable," said Scott. "We have climbed a steep hill and we didn't even really realize it. And we're going like 10 clicks (10 kilometers an hour) up this hill, a slope of 8 degrees or so. And we can look back and see the LM just as loud and clear as can be. Joe, when the TV comes on, you're going to get a super picture."

The leaning lunar module was 4½ miles away. Beyond it curved the arcuate shoulder of the Apennines and above them the Moon's black sky. The view was stunning. Later, Scott recalled: "Taking it all in, our landing craft, the rolling plain and the mountains, I began to feel at home in our new surroundings. Yes, I actually felt at home and couldn't help thinking to myself: Gee, everything is working. Here we are in this beautiful place and every time we turn around we find something new and extraordinary."[10]

Irwin advised Mission Control to aim the TV camera toward Mt. Hadley about 12 miles away and see if it would pick up the linear pattern that seemed clear to the explorers. Meanwhile, Scott and Irwin dismounted from the Rover to collect rocks and take a core sample. Several times, Houston heard Scott calling out: "Watch out, Jim. Watch out!" He was warning Irwin not to step back from a ledge.

After several minutes of prospecting, Scott saw rocks that looked green in color. He photographed a huge block that was 3 meters (10 feet) long. It had a rough surface, with a 2-foot layer that appeared absolutely green. Scott could hardly believe it. People had kidded him about finding green cheese on the Moon, but green rocks! He wondered if it was an illusion.

Irwin found a 3-inch sample of the green rock and put it into one of the sample bags. Scott examined a boulder ensconced on the side of Spur Crater. When he scraped the outer crust away, a greenish gray material appeared, similar to the green rock fragments they first saw on the ground.

The Rover, meanwhile, was parked perpendicular to the steep slope so that it would not roll downhill, but it began to slide sideways until Irwin grabbed it. He held the machine until Scott collected a greenish sample from the boulder and took pictures. Then the two of them climbed into the Rover and drove slowly down the slope, stopping once more to collect rocks.

Again, they encountered the green rocks. Scott still couldn't believe the rocks really were green. It must be some optical effect, he said.

"Oh, it's a good story, talking about green cheese," said Irwin. "Who would ever believe it? I hope it's green when we get it home."

They examined the new find carefully.

"Oh, my. It is green. It is green," said Scott.

"I told you," said Irwin. "It's green."

"Fantastic," said Scott. "This has got to be something. And it looks (he raised his sun visor) . . . hey, now it's gray! The visor makes it green, Jim."

Later examination did confirm that the green rocks actually were somewhat green—although not as green as they appeared to be when viewed through the helmet sun visor. The green rock was a kind of clump consisting of compacted soil and rounded glass spherules. The samples were as crumbly as clods and in the laboratory the spherules had the color of an empty Coca-Cola bottle.[11]

Another Genesis Rock

Near the rim of Spur Crater they found a large breccia sitting on a clod pedestal, waiting for their attention. Embedded in the breccia was a clast of anorthosite with plagioclase crystals 2 centimeters long. It was the largest piece of anorthosite yet found on the Moon and the explorers concluded that it was the "genesis" rock they were seeking. At least, it was different than any other rocks they had seen and possibly represented a piece of the original crust.

The sample (No. 15415) turned out to be one of the oldest rocks found on the Moon. It was dated by a team at the State University of New York, Stony Brook, at 4.09 billion years plus or minus 100 million years by the potassium–argon method.[12]

Other bits and pieces of anorthosite had turned up in samples returned from Mare Tranquillitatis, Oceanus Procellarum, and Fra Mauro. The first clue that these ancient pieces were scattered over the Moon was "Luny Rock 1," an anomalous fragment returned by Apollo 11. But it was questionable whether they would qualify as parts of the original crust. The primitive crust would be nearer 4.6 billion years old, like the ancient component of the soil that kept showing up on mission after mission—the component that was older than the rocks it supposedly came from.

"Joe, this crater is a gold mine," said Scott, delighted with the finding of the anorthosite.

"And there might be diamonds in the next one," said Allen. "Dave, we're coming up on departure time. All we really need is soil from this same area. And we're making money hand over fist."

Scott and Irwin bagged the soil "the boys in the back room" wanted and climbed into the Rover, experiencing the usual trouble fastening seat belts. Ahead of them, Mt. Hadley was now highly illuminated. The lineations of banding that sloped across the face were spectacular. As Allen stated later, the array of "stark, sharply etched, parallel linear patters was completely unexpected."[13]

Whatever process was indicated by this banding was not clear. Allen suggested later that the major lineaments might represent compositional layers or regional fractures, but the ambiguities introduced by the oblique lighting made an unequivocal interpretation difficult.

On Earth such banding could readily be interpreted as representing periods of deposition of sediments in water and the subsequent uplifting of sedimentary rocks by mountain-raising forces. On the Moon sedimentation had to be explained in terms of accretion of material from impact ejecta or dust fall. The lineaments on Mt. Hadley seemed to be uniformly about 35 feet thick.

On the trip back to Hadley Base, the explorers were allowed a 15-minute stop at Dune Crater, one of the largest and most intriguing to them in the whole area. They collected rocks and departed on time, but it was difficult to leave the place. In fact, it was difficult, they said, not to stop every several feet and pick up something that appeared to be new, so great was the rock variety in that region. When they reached base, the Rover odometer showed they had traveled 12.7 kilometers, or 7.6 miles.

The second traverse had been observed at Houston by a number of VIPs, including John Glenn, then a civilian businessman with political ambitions; Edward E. David, President Nixon's science advisor and the last scientist to serve in that capacity before the President relegated the science advisory function to the National Science Foundation in 1973; NASA Administrator James C. Fletcher, the former president of the University of Utah, who succeeded Paine in 1971; and Fletcher's deputy, George M. Low.

There was unfinished work to be done on ALSEP and the explorers tackled it without delay. Scott took up the jackhammer drill and tried to deepen the holes for heat flow probes. But the drill would go down only 1.6 meters (4.2 feet) at one hole and a little more than 1 meter at the other. Because of a break in the drill stem, one of the probes could not be pushed all the way down. Although this was considerably short of the 10-foot depth the experimenters wanted, it was the best the lunar laborers could do.

The astronauts found that resistance of the regolith was greater than expected. The drill stem did not clear dirt out of the hole effectively and became stuck in the drill chuck. However, they succeeded in emplacing the probes in the shallow hole and hooked them up to the electronics box, which fed data to the radio transmitter.[14]

When this task was completed, they fitted a core tube into the drill and tried to bore 10 feet down to extract a deep core. They managed to penetrate 2.4 meters (7.8 feet) with the core tube, but found there was not enough time left in the EVA to pull out the core—which required a considerable amount of effort. They left this chore for the next day.

Irwin carried out a soil mechanics experiment that had implications for future operations on the Moon. He dug a shallow trench, using a small scoop attached to an extension handle. He managed to dig down without difficulty to a depth of 41 centimeters (16 inches) when he hit "hardpan." Photographs of the trench and its walls provided clues to the mechanical properties of the soil. Someday this information might be useful to a future expedition seeking to erect a shelter. At the bottom of the trench, Irwin tested resistance to further penetration with a pole-like instrument called a penetrometer. It measured the amount of force required to penetrate the soil at various depths.

The last task was to erect the United States flag. Irwin pushed the staff into the ground and hit it a few times with a hammer "so it will stay up here a few million years." It was an interesting idea to contemplate. If the insignia of Rome had been planted on the Moon in Caesar's time, it would still be there, virtually unchanged. From the viewpoint of man's time on Earth, the flags at the Apollo landing sites, and the equipment, would remain as markers to the first human arrivals over periods of geologic time.

The day's work over, the explorers climbed back into the LM for a meal of processed food, fruit drink, cocoa, and 6 hours of rest. They did not feel especially tired, although it had been a long and active day. They had spent 7 hours, 12 minutes, and 46 seconds outside, a new EVA record.

Late Sunday afternoon, the Endeavour came over the site and Irwin called up to Al Worden: "Hey, Al, throw down my soap."

Worden replied, "Forget something, Jim?"

Irwin said, "I really need my soap."

The exchange meant was that all was going well.

Post Office

Scott and Irwin awoke shortly after midnight. In the first hour of Monday morning, Houston time, they were ready for their third traverse on the Moon. Houston had awarded them an extra 90 minutes of rest time and this shortened their third EVA. The passive seismometer was operating; it recorded their movements in the LM cabin and their climb down the ladder to the ground.

"Nice to be outside where you can stretch a bit," said Scott. He and Irwin ambled, kangaroo style, to the core tube that they had left in the soil during their rest period. It took their combined strength to pull the tube out of the ground. The lowest section broke off. But they had a core nearly 7 feet long, the deepest sample yet extracted from the Moon.

Allen suggested that they take "a good, clean, comfortable look at the rille" and away they went in the Rover. At 5:47 A.M., Worden, passing overhead, called down that he had changed the Endeavour's orbital plane in preparation for rendezvous and docking with the Falcon later in the day. Scott and Irwin sent up a "thank you" and continued driving westward.

They parked on the rim of the rille and Scott described the scene to Houston. From the top of the rille down, he saw debris all the way. Horizontal layering appeared near the top. It extended downward about 10 percent of the way, he said. Below was one layer that extended about 40 percent of the rille depth.

After their return to Houston, the explorers said they believed the layering represented a series of basalt flows. The 10-meter blocks they saw elsewhere indicated very thick and cohesive flows.

Alighting from the Rover, the explorers eased part way down the rille slope, picking up rock fragments with the tongs. Once Irwin stumbled but then regained his balance. They chipped off pieces of half-buried rock that might be bedrock. Layering seemed to be a characteristic of the larger rocks and boulders of the region. Lava flows seemed to be another explanation. The mountains were layered. The rille walls were layered. They found a layered boulder 1 meter long. After taking a core sample, they drove back east toward the LM. Mt. Hadley was brilliant in full sunshine.

Scott said, "Oh, look at the mountains today, Jim, when they're all sunlit. Isn't that beautiful?"

Irwin answered, "Dave, I'm reminded about my favorite biblical passage from Psalms. I'll look unto the hills from whence cometh my help. But, of course, we get quite a bit from Houston, too."

They picked up their outbound tracks and shortly thereafter Scott reported they were "back on the friendly plains of Hadley."

Parking the Rover, Scott performed a chore that seemed innocent enough at the time but had embarrassing repercussions later.

"To show that our good postal service delivers any place in the universe," he announced before the TV camera, "I have the pleasant task of cancelling here on the Moon the first stamp of a new issue dedicated to commemorate United States achievements in space. I cancel this stamp August 2, 1971, the first day of issue. What can be a better place to cancel a stamp than right here, the Hadley Rille?"

"Now," Irwin added, "I'll stick this back in a special mail pouch here and we'll deliver it when we return."

When the ceremony ended, Scott demonstrated Galileo's famous experiment on gravitational attraction, showing that objects of different sizes and weights fall with equal velocity. Galileo is supposed to have dropped stones of different sizes from the Leaning Tower of Pisa. Scott dropped a feather he had brought for the occasion and a hammer in front of the camera. He then announced the result: "This proves that Mr. Galileo was correct."

Then, in a final peroration, he said: "I think that's pretty good after only 10 years. Here we are spending 3 days on the Moon. That's moving ahead."

The EVA was cut to 4 hours and 50 minutes to meet the liftoff time for the return to the Endeavour. For the first time, the launch of the ascent stage of a lunar module from the Moon was observed on television—by the remotely controlled camera on the Rover.

The countdown was heard from Mission Control and suddenly there was a puff of dust and the ascent stage went up like a bullet. There was none of the lingering so characteristic of an Earth launch. Falcon vanished in a few seconds, not only because of its velocity but also because the elevation clutch on the camera slipped and the lens could not be elevated to follow the departing bird. The camera was turned off. About 40 minutes later, it was turned on again to survey the scene, with the descent stage standing there, the Rover, the flag, and miscellaneous junk littering the surface. The camera worked 13 minutes, and then all went dark.

In orbit aboard the Endeavour, Worden had used a metric camera to photograph the surface for a map with a resolution of 25 meters. Superimposed on each frame was the altitude of the spacecraft, although the pilot had difficulty with the laser altimeter provided to give altitude. The location of each frame was pinpointed by a concurrent star photograph made by a 35-millimeter camera pointing toward the star field. Worden estimated that 10 percent of the lunar surface was photographed from the Endeavour.

While his colleagues were on the surface, Worden had studied the large feature geology beneath his flight path. He had been instructed to pay particular attention to the Craters Tsiolkovsky, Proclus, and Aristarchus and the Littrow Crater on the eastern edge of Mare Serenitatis. The Littrow Crater area was being considered as the site of Apollo 17.

Worden described the far side of the Moon as "very hummocky" with subdued craters and few distinctive lava flows or mare areas. Basin-type features did not appear until the vessel approached the Crater Tsiolkovsky, he said.[15]

The laser altimeter showed that lunar far side was 5 to 10 kilometers higher than near side. A fault zone appeared to run through Tsiolkovsky from north to south, and parallel lineaments extended 15 to 20 degrees west of the crater rim. There was evidence of a rock avalanche from the rim of Tsiolkovsky, Worden reported, and in the crater's central peak he could see layering similar to that on Mt. Hadley.

On the eastern side of Mare Serenitatis, Worden saw dark deposits around the Littrow Crater. They seemed to be associated with arcuate rilles and with cinder cones. The cinder cones were evidence of probable volcanism. One photographic enlargement showed not only cinder cones but halos around cones and craters, indicating recent volcanism, Worden said. He said the area was passable for a landing.

Worden said he saw structures that he interpreted as volcanic on the plateau around Aristarchus. The rilles there appeared to be older than the mare surface because they seemed to have lava in them.

On the afternoon of August 7, 1971, Scott, Irwin, and Worden splashed down in the Pacific Ocean within 6 miles of the U.S.S. *Okinawa*. They brought with them 77 kilograms (169.4 pounds) of rocks and soil and a whole library of photographic film unmatched on previous flights.

They had nearly doubled lunar surface time from 34 hours on Apollo 14 to 66 hours, 54 minutes; they had roamed the surface for 18 hours and 30 minutes, about double the Apollo 14 EVA time; and they had introduced the first wheeled spacecraft to the surface of another planet.

Notes

1. Berry, Charles, M.D., Medical Results of Apollo 14, NASA, presented at the 22nd International Astronautics Congress, Brussels, September 25, 1971.
2. Ibid.
3. Berry, Charles, M.D., Biomedical Findings on American Astronauts Participating in Space Missions, 4th International Symposium on Basic Environmental Problems of Man in Space, Yerevan, Armenia, USSR.
4. Ibid.
5. NASA Budget Plan Summaries, 1971 and 1973.
6. Ibid.
7. Latham, Gary, NASA Science Briefing, Houston, May 26, 1971.
8. Baldwin, Ralph B., *The Measure of the Moon*. University of Chicago Press, Chicago, 1963; and Kozyrev, "Observations of a Volcanic Process on the Moon," *Sky and Telescope*, Vol. 18, 184 (1959).
9. Personal account, *The New York Times*, August 14, 1971.
10. Personal account, *The New York Times*, August 13, 1971.

11. Pilots' Report to the Lunar Science Conference, Houston, January 12, 1972.

12. Schaeffer, O. A., et al., "Ages of Lunar Materials from the Hadley Rille Area . . . ," State University of New York, Stony Brook, Third Lunar Science Conference, 1972.

13. Allen, Joseph P., Summary of Scientific Results, Preliminary Science Report, NASA.

14. Crew Observations, Preliminary Science Report, Apollo 15, NASA.

15. Ibid.

10

The New Moon

The flight of Apollo 15 was the high watermark of lunar exploration during the voyages of Apollo. With a grand total of 387 pounds of lunar rocks and soils from four missions, seismic data from three stations that continued to "listen" for impacts and quakes, new magnetic evidence, new heat flow data, and the results of X-ray, gamma-ray, and photographic mapping from orbit, a new picture of the Moon had emerged by the beginning of 1972.

A structure had appeared. The "new moon" had a crust 65 kilometers (39 miles) thick in the Fra Mauro region, 10 times thicker than the Earth's crust in the oceans and of similar thickness to the Earth's crust under mountain ranges. Below that, there was a region analogous to the Earth's mantle and, if one accepted the magnetic evidence, a small core, possibly of iron.

The Moon was a planet more like the Earth than like the meteorites, the only other extraterrestrial objects man had been able to analyze. Yet there were chemical distinctions that tended to negate any parental relations. The ratio of potassium to uranium was much lower than on Earth. The Moon was depleted of volatile elements. It had a dearth of oxygen, no water, and, consequently, was denied the chemistry of life as it proceeded on Earth.

The predominant view of the new moon arising from the Apollo discoveries by early 1972 held that the Moon had undergone several periods of heating. These had produced the thick, layered crust. In the first 300 to 400 million years after its formation, the Moon had acquired a thin, low density scum of anorthosite that had been smashed and virtually pulverized by the bombardment of planetoids, asteroids, meteorites, and comets.[1]

During this early period, the mare basins were formed by gigantic impacts. Successive flows of lava poured into them and basalts of two types were generated. One was the iron and titanium-rich mare basalt found at Tranquility Base. The other was a basalt with a higher concentration of radioactive elements called KREEP, an acronym describing its abundance of potassium (K), rare earth elements and phosphorus (P). It was found in oceanus procellarum. The iron and titanium content of the mare basalts made them appear dark. The bright regions of the Moon, the highlands or terrae, were characterized by the lighter colored, calcium–aluminum silicate rock called anorthosite, which is low in radioactive elements and iron. There were thus three main rock provinces on the Moon, plus an occasional piece of material that did not seem to fit any category.

It seemed to be clear enough that the only way this diversity of rock types could be accounted for was by accepting a heterogenous Moon at the outset. In the predominant view, there was little doubt left by 1972 that here was a planet with an early superficial crust and a history of partial melting, going deeper and deeper with time and producing lavas of varied composition. Then, about 3.1 billion years ago, the lunar heat engine shut off. No rocks with confirmed radiogenic ages younger than that were found; presumably none had crystallized after that time. Of the rocks found at four Apollo landings and the site of the Soviet Union's Luna 16 grab sample, none was older than 4.2 billion years nor younger than 3.1 billion years. The soils were older than the rocks and dated back to 4.5 or 4.6 billion years. There were ancient materials in the breccias 4.4 billion years old. But as far as the rocks returned to Earth showed, igneous processes on the Moon continued for 1.1 billion years and stopped.

Lunar volcanism had ended, except for small, local pools of residual magma that were suggested by mild Moonquakes and intermittent emissions of gases from craters such as Aristarchus and Alphonsus. From the evidence accumulated from Apollo 11 through Apollo 15, the Moon had been volcanically senile for 3.1 billion years, but it was not dead. The Moonquake evidence strongly hinted at deep magma pools scattered beneath the crust. At least one was believed to be a kilometer or two in diameter.

The existence of such hot spots was supported by data from the heat flow experiment that Scott and Irwin had struggled so hard to emplace. Over a period of 45 days, the average heat flow through the regolith at the Apennine Mountain front was 3.3 millionths of a watt (3.3×10^{-6} watt) per square centimeter. This value is about half of the average heat flow of the Earth, the experimenters said. They surmised that most of the present heat flow comes from the decay of long-lived radioisotopes.[2]

During the last billion years, however, the lunar surface had been sculpted and gardened by the infall of asteroids, meteorites, and comets in Sunbound orbits. The bombardment was tailing off then. In the beginning, the Moon had been struck by massive objects, possibly planetesimals that had never become planets. One of the last of these was a body believed to be the size of the Island of Cyprus that had gouged out the Imbrium Basin 3.9 billion years ago, judging from the exposure ages and formation ages of rocks returned by Apollo 14 and 15.

Big craters were punched in the surface for a long time after that, but with Imbrium, the basin-making period of the Moon's history apparently ended. The Crater Copernicus, dated by the material found in its rays, was 850 million years old, relatively recent for the Moon.

These views seemed to be predominant when the Third Lunar Science Conference convened at the Manned Spacecraft Center, Houston, January 10–13, 1972. However, there were strong minority opinions. They challenged the conclusions of widespread melting and of the existence of core, about which there was sharp dispute.

The divergence in ages between rocks and soils was not entirely resolved. The existence of some very ancient component had to be assumed, but—what was it? The dating game the scientists played with the lunar rocks was characterized by discrepancy, inexactness, and contradictory measurements. Chronologies varied with the laboratories performing the analyses. There had been persistent evidence that some portions of the soil were older than the meteorites. The meteorites are the Rosetta stones for the Solar System. They are dated at 4.5 billion years. Since they are composed of primitive material, they are believed to be the earliest condensates from the solar nebula. Thus, in the light of this belief, the Moon could not be older than the meteorites—otherwise, cosmogonic chronology was askew.

In general, Leon Silver explained, the ages obtained by measuring the decay of uranium and thorium to lead are older than those obtained by the potassium–argon or rubidium–strontium decay methods. Unless there was error, this implied that the uranium–lead system data were revealing materials more primitive than the rocks they formed—perhaps materials that had coalesced in the solar nebula.

"The most popular way of explaining it," Silver said, "would be to suggest that the isotope systems of uranium, thorium and lead have a memory of some prior stage of rock evolution."

The question of the origin of the Moon was still an open one at the time of the Third Lunar Science Conference. The fission theory had been weakened. The Saturnian-ring theory looked a little better than before. The capture theory had been modified by partial agreement that the Moon could not have been captured later than 4 billion years ago.

One main point had been demonstrated beyond dispute. There was no life on the Moon, and there never had been before man's arrival. Thus, the lunar environment represented one boundary condition for the evolution of life. Living systems could not develop on a waterless planet as deficient in volatiles as the Moon. But there were traces of compounds that are recognized as the precursors to those that evolve to become living systems in more favorable environments. Similar materials for the building blocks of life are found in a class of meteorites called carbonaceous chondrites. It appeared a priori they were a part of the solar nebula from which the Sun and the planets formed. If so, life might arise anywhere conditions made it possible.

The Crust

The identification of a layered crust 65 kilometers deep at the Fra Mauro region of Oceanus Procellarum was one of the major new findings reported at the Third Lunar Science Conference. Seismic data had shown a layer of basalt some 25 kilometers thick and an underlying layer of anorthositic gabbro 40 kilometers thick.

The KREEP basalt was curiously localized. It was virtually confined to the western maria, Procellarum and Imbrium, although bits and pieces of it had been picked up by Armstrong and Aldrin at Tranquility Base amid the dominant iron–titanium-rich mare basalt, which was more generally distributed.

To the geologists and geochemists, the existence of these two families of basalts was evidence of extensive melting in the Moon during the period of mare basin filling. The discovery of anorthosite below the basalt was an indication of a volcanic period preceding the basin-filling period.

Thus, the wiggles and squiggles on the seismograms at Houston showed that the lunar crust was a piece of cake, a layer cake, with the main anorthosite–norite body covered with a dark layer of basalt—at least in the Fra Mauro region.

This picture had been drawn mainly from the seismic signals of six artificial impacts—three of the lunar module ascent stages on Apollo 12, 14, and 15 and three S4B drops from the launches of Apollo 13, 14, and 15. The Apollo 13 LM ascent stage was not dropped because it was needed as a lifeboat to bring the crew home.

For the geologists the existence of a thick crust confirmed the mag-

matic process, which was the way they preferred to explain the igneous
rocks. There was another way of accounting for basalts by hypothecat-
ing that they had been differentiated before the Moon was formed, but
geologists would not accept this idea.

A thick crust indicates a "very, very extensive melting in at least the
outer region of the Moon," said Gary Latham.[3] A residue of the melt
still existed. Analyses of 1800 Moonquakes recorded by the seismometers in Oceanus Procellarum, Fra Mauro, and the Palus Putredinus–
Apennine region on the borders of Imbrium pointed to the existence of
a lake of magma as a probable Moonquake source at a depth of 800
kilometers (480 miles).

During May 1971, Latham said, a series of small quakes gradually
increasing in amplitude for several days was noticed. The episode ended
with a big Moonquake—the largest recorded up to that time. This kind
of quake activity has been seen on Earth in volcanic regions, Latham
said. As the hot magma migrates upward, it produces a series of disturbances at depth and these become stronger as they continue. Finally,
when the volcano erupts, it is heard as the strongest of the quake series.

Did this mean there was a live volcano actually erupting on the
Moon? No one had seen any evidence of it during May 1971. The best
location the seismology team could deduce for the event was in the
Mare Crisium (Sea of Crises) about 1300 kilometers (780 miles) east
of the Apollo 15 seismometer.

The evidence was tantalizing. Although the main episodes of lunar
volcanism ceased 3 billion years ago, Latham observed, volcanism
could be continuing on the Moon on a minor scale.

Evidence of extensive volcanism in the past was amplified pictorially
in the photographs Worden took while orbiting in Apollo 15. From
these it was possible to trace the full extent of a mare basin and see
successive lava flows.

A geologist with Bellcomm, Farouk El Baz, said that the photos
"give you a feel for how extensive they were and how long they persisted."[4]

The Cornell astronomer Thomas Gold uttered a strong dissent. Before the landings on the Moon, Gold had hypothesized that the surface
was composed of deep dust. It was considered questionable that it
would support the weight of a spacecraft or a man until the Surveyors
landed successfully on both mare and highland material. Gold held to
the dust hypothesis. It explained at least the powdery regolith, the top
soil of the Moon. "I never cease to be amazed at the degree of enthusi-
asm that the volcanists have for their processes," he said.[5] "Which
processes, if true, are very interesting and fascinating on the Earth. But

the hope of finding them on the Moon at the present time is absolutely wasted motion, because there are some 900 volcanoes on the Earth that spew out each week more stuff than we would readily see anywhere on the front side of the Moon if it were spewed out. You know there is not a single volcano on the Moon which is as active as the first 900 are on Earth."

The sighting of luminous hazes and mists on the lunar surface for centuries was explained by Gold as probably the emission of a few tens of kilograms of gas. "It is quite a minor volcanic event on Earth which spews out a million tons," he said.

Gold offered an alternative explanation to volcanism as the cause of the layering of the crust. He suggested that the layers represented build-up by accretion.

First of all, he maintained, the Moon was not formed in a single accretionary episode. There may have been several intermediate accumulations that broke up and continued accreting on their own until they once more coalesced to form the final body. The intermediate bodies would have undergone the cooking and differentiation attributed to the whole Moon from the heat released by short-lived radioactive isotopes, now decayed, so that when the final body was formed, the differentiated material, the basalts and the anorthosite, were already there—precooked, so to speak.

The Moon had not formed from a primitive mixture, Gold insisted, but from one already processed. Thus, the rocks that had the appearance of having crystallized on the Moon from a liquid might actually have been formed in a more primitive body and were later deposited on the Moon by the infall of planetesimals. Perhaps the igneous materials found on Earth had not formed there either but had been acquired by accretion, Gold suggested. If so, he said, "a lot of geology would have to be revised."

The geologists, of course, would not buy any part of Gold's theory. But the conflict between geologists and astronomers was nothing new; it had been going on since the days of Hutton and Lyell. It had been simply transposed from the Earth to the Moon.

Gold also disputed the theory of Ringwood, the Australian physicist, that the Moon had coalesced from a primitive atmosphere of hot silicates orbiting the Earth. Such a process, said Gold, was dynamically impossible.

To those who maintained that the whole Moon had melted, not merely its outer regions, Gold offered this argument: If the interior had been molten, it would have "washed out" its internal lumps of high density material that are believed to account for its asymmetrical shape and its displaced center of mass. The existence of such inhomogeneities rules out a molten interior, Gold said.

The geologists, he said, had an "unbalanced" point of view because "they keep searching as hard as they can for a terrestrial-like process." But the Earth analogy does not hold.

For example, Gold pointed to the rounded, dunelike mountains. Everything on the Moon was rounded, all covered with a layer of powder "at least some tens of feet thick," he said. Geologists kept looking for bedrock on the Moon, a chimerical notion, in Gold's opinion. There was no bedrock on the Moon, he insisted.

This view was challenged by the Princeton University geologist Robert Phinney, who said that one of the slides Gold had used to illustrate his lecture at the Third Lunar Science Conference showed bedrock clearly. Photos of the Hadley Rille taken by Scott and Irwin showed bedrock at shallow depth under the rubble, Phinney maintained. "One of the important outcomes of the last two missions has been the clear demonstration that the methods of geology are applicable to the surface of the Moon," Phinney said.[6]

The differences in lunar and Earth geology, though, were striking. On the Moon a period of 300 million years, which is more than half of known geologic time on Earth, is represented by granular material a foot deep. In that span of time on Earth sediments kilometers thick have been laid down by running water, Phinney noted.

Compared with the Earth, the lava outpourings on the Moon are quite remarkable. One comparable terrestrial outpouring is the Columbia River Plateau. The volume of lava there is comparable to some of the smaller mare surfaces on the Moon, observed Brian Skinner, a Yale University geophysicist.

Central Heating

The source of the heating that had formed the igneous rocks was argued rather hotly at the Third Lunar Science Conference. Experts were divided not only about its extent but whether it was volcanic or gravitational, the product of radioactive decay or of impact. The heat flow experiment on Apollo 15 supported volcanism, as did the preponderance of the evidence, but the arguments went on.

Wasserburg defined the problem at a news conference. There was a tendency, he said, to propose "a half-baked Moon model, namely, one heated from the outside, baked a little bit from the outside, and melted partly." Magnetometer evidence indicated, however, that the Moon had been baked all the way through. Magnetically, it was well done.

This conclusion was based on remanent magnetism found at the Apollo 12 site of 38 gammas, at two points on the Fra Mauro formation where readings of 103 gammas were made near the LM and of 43 gammas about a kilometer away, and at the Hadley-Apennines where readings were 6 to 10 gammas. These results were reported by a team composed of C. P. Sonnett, principal investigator, Palmer Dyal of NASA's Ames Research Center, and C. W. Parkin, a postdoctoral associate at the National Research Council.

An English geophysicist, S. Keith Runcorn of the University of Newcastle upon Tyne, interpreted these readings as implying the existence of a Moon-wide field of at least 1000 gammas in the past. This is only about $\frac{1}{30}$ of the Earth's magnetic field at the equator, but its implications are interesting and Runcorn was not hesitant about spelling them out. He concluded that "the Moon possessed a field of internal origin from its early history until some time no later than 3.2 billion years ago."

An internal field, according to present theory, requires an internal dynamo producing electrical currents. That requires an iron core. Runcorn hypothesized that 500 million years after the Moon formed, temperatures at a few hundred kilometers' depth rose to 900°C. Partial melting ensued, resulting in the formation of the anorthositic highland crust. After this shell cooled, it was punctured by planetesimals that formed the circular maria. Then, some 500 million years later, the basins filled with lava welling up through cracks in the shell from a region 200 and 300 kilometers deep. By that time, the interior had reached a temperature of 1200°C at which basaltic lava would flow, Runcorn held. At the center of the Moon, an iron core formed, he theorized. It was perhaps about two-tenths of the Moon's radius (216 miles) in diameter. This core could have produced the magnetic field during the time it was molten and at a time when the Moon was rotating faster than it now does. Runcorn ruled out any possibility that the Moon could have been magnetized externally, either by being closer to the Earth or by the solar magnetic field.

Critics of the hot Moon theory, as reported earlier, had insisted that the Moon's well-known deviation from hydrostatic equilibrium ruled out any likelihood that it had melted all the way through. A molten Moon would have assumed the shape and mass distribution of a perfect sphere, they contended.

Runcorn countered this objection by asserting that the Moon's departure from hydrostatic equilibrium was the result of thermal convection, arising from its early high temperature. As the Moon cooled and convection ceased, the deviation was frozen into the lunar structure.

Thermal convection was the only way the Moon could have cooled, Runcorn added.

Summarizing his ideas, Runcorn said:

> I think that it is quite clear that the small, electrically conducting core is necessary. You've got to have an electrically conducting region, a region of metallic conductivity. Now for the terrestrial planets, geochemically, this must mean iron. When you're talking about the magnetic field of Jupiter, we know there that the likely conductor is metallic hydrogen . . . but that doesn't concern us in the terrestrial planets.
>
> The reason people find an iron core in the Moon difficult to accept is that, for years, you see, we have thought that because of the density of the Moon, 3.34 (grams per cubic centimeter), very near that of olivine, there was no room at all for an iron core. But an iron core of about one-fifth of the radius is compatible with all we know now about the density and moment of inertia.[7]

The View from Lunar Orbit

One of the important clues to a heterogenous Moon was a rock, No. 12013, picked up at the Apollo 12 site in Oceanus Procellarum. It was 13.75 centimeters (5.36 inches) long and about half that in width, tapering at each end like a crude stone axe. About 70 percent of it was plagioclase. It was not related to the basaltic rocks among which it was found. It contained the highest amount of silicon dioxide seen in any lunar sample up to that time and was enriched in potassium, rubidium, barium zirconium, yttrium, lithium, and the rare Earth element, ytterbium, 10 to 50 times as compared with other rocks. From this sample alone, it was evident as early as 1969 that the Moon was a more complex and heterogenous body than many scientists had supposed. The question this and the KREEP-type rocks raised was: What extent were various materials distributed over the surface of the Moon?

Apollo 15 provided a means of finding out which did not involve wholesale trekking over the surface. The X-ray and gamma-ray detectors that the Endeavour carried were designed to map the broad-scale chemistry of the surface to the extent of revealing chemical provinces. The X-ray fluorescence detector registered the common elements aluminum, silicon, and magnesium, which were the principal constituents of the light-colored anorthosite and seemed to be predominant in the highlands. The gamma-ray spectrometer detected the emission of gamma radiation, the signature of radioactive elements—uranium, thorium, and potassium. This instrument would pick up areas of

KREEP-rich rocks and soils. The third detector on board, an alpha-particle spectrometer, sensed radon, an unstable, intermediate product of the decay of uranium and thorium. Radon emits an alpha particle as it decays further into polonium. Emission of an alpha particle (helium nucleus) not only points to areas of radioactivity but indicates out-gassing from volcanic activity, even at a very low level. We will hear more of this experiment later.

Meanwhile, the X-ray and gamma-ray detectors drew a chemical map of the surface under the flight path of Endeavour. The X-ray fluorescence spectrometer detected secondary X-ray radiation emitted by aluminum, silicon, and magnesium when they were fluoresced by solar X rays. These elements were the principal constituents of the light-colored anorthosite.

The Apollo 15 ground track covered 15 percent of the surface. Only half of that was in sunlight. Consequently, the X-ray fluorescence detector, which functioned only in sunlight, was able to "see" about 7.5 percent of the surface. But even that was a large area.

The principal investigator, Isidor Adler, a senior scientist at NASA's Goddard Space Flight Center, and his colleagues were able to plot a map along the near side ground track showing the abundance of these elements and their variation across the face of the Moon. The results proved that the highlands are different in chemical composition from the maria, as anyone might suspect from simply looking at them. The highlands are bright because their aluminum–silicon composition reflects more light than the dark iron-rich maria.

"The high aluminum–silicon concentration ratios of the highlands suggests they are related to the anorthositic–noritic fractions of the returned samples," Adler reported.[8] "While the low aluminum–silicon concentration ratio of the maria is consistent with the composition of the mare basalts analyzed."

Thus far, there were no surprises. But the gamma-ray experiment revealed one. It showed that radioactive elements (uranium, potassium, and thorium) were concentrated in one region of the western maria—in Imbrium and in the great lava flows of Oceanus Procellarum, where the KREEP basalts were found.

On the far side of the Moon, 180 degrees away, near the Crater Van de Graaff, the gamma-ray spectrometer picked up a smaller concentration of radioactive minerals. Unlike the X-ray detector, the gamma-ray device was not limited to the sunlit portion of the Moon. The concentrations of radioactive elements in the region of the western maria were similar to those of the KREEP basalts, according to James R. Arnold of the University of California, San Diego, the principal investigator on the gamma-ray experiment.

Arnold and his colleagues were intrigued by the concentration of these elements in one major and one minor region. What process had brought this about? Was it a product of differentiation on the Moon? Had these elements been concentrated during accretion? Was it analogous to the formation of ore bodies on Earth?

The finding, however, might not have been completely unexpected. It was suggested by the samples that Shepard and Mitchell returned from the Fra Mauro formation. Some of these were higher in potassium, thorium, and uranium than those brought back from Tranquillitatis or Procellarum. One basaltic rock, No. 14310, contained the highest amount of these elements of any rock found on the Moon up to that time. The sample had 3 times as much thorium and uranium as most terrestrial basalts.[9]

Beyond the main area of high radioactivity, Arnold reported, "what you see for 800 to 900 kilometers on either side is a gradual falling off. Over the eastern maria, like Crisium and Smythii, we see no measurable rise in radioactivity at all. Only on the southern back side do we see a slight rise in radioactivity—about 20 percent as high as the Imbrium material."[10]

The Assessment

It was now possible to assess the general distribution of rock types with the evidence of these experiments and the analyses of samples returned by four Apollo landing explorations and the Soviet Union's Luna 16 grab sample.

As the Lunar Sample Analysis Planning Team summarized the results of the five missions, the most significant gains in knowledge about the Moon were:

We now know there is a lunar crust; we know the principal types of rock that comprise it and, broadly speaking, their distribution over the surface of the Moon. We know the thickness of the crust in one region.

There are three principal classes of crustal rock. One is mare basalt, rich in iron, and sometimes titanium. The Tranquility Base rocks are examples. Another is noritic rock, rich in radioactive elements and refractory trace elements (KREEP). The third is aluminum-rich, anorthositic rocks. The most abundant material of this class appears to be anorthositic gabbro, containing 70 percent plagioclase, although rocks with greater plagioclase content are also present.[11]

The importance of this statement is that for the first time, the lunar surface was organized into a meaningful pattern that seemed to reflect its evolution. Before this pattern was realized fragments and bits and pieces of all three of the main types of igneous rocks, scattered at various sites and compacted in the breccia, presented a chaotic picture of the surface. Armstrong and Aldrin had brought back an anomalous rock containing material from all three types, the breccia called Luny Rock 1. It contained not only KREEP but also mare basalt and anorthositic plagioclase—all three mashed together by ancient impacts.

When Luny Rock 1 was first examined, no one could be sure what it signified. By the time of the Third Lunar Science Conference, the enigma of Luny Rock 1 was resolved. It was simply a composite of the lunar surface.

Similarly, although anorthosite fragments had been found in breccias from Tranquillitatis, Procellarum, and Fra Mauro, the first large specimen of lunar anorthosite to be collected was picked up in the Apennines by Scott and Irwin. This was rock No. 14,515, which they found on the north-northwest rim of Spur Crater. It was so-called "genesis rock"—at least the genesis rock of that mission.

It was dated by the argon-40–argon-39 method by Oliver A. Schaeffer, Liaquat Husain, and John F. Sutter at the State University of New York, Stony Brook. They reported that its age was 4.09 billion years, plus or minus 190 million years. They concluded this age was close to the time when "the anorthosite complex was formed near the lunar surface." They concluded also that "it has been virtually unaffected by postcrystallization, thermal events." Thus, they said, "the lunar anorthosite is probably a fragment from the primitive lunar crust which was formed about 4.1 billion years ago."[12]

Another rock dated by the Stony Brook laboratory was a basalt that Scott and Irwin returned from the Hadley Rille rim (Rock No. 15,555). Its age appeared to be 3.28 billion years, plus or minus 6 million. It was taken to represent a late stage of lava production in the Imbrium Basin.

Conceding that the data still were limited, especially from the lunar highlands, the Schaeffer group proposed this chronology of lunar evolution: The crust, or at least part of it, was formed 4.1 billion years ago and the Imbrium Basin about 300 million years later, judging from the dating of rocks found in the Fra Mauro formation. Mare Tranquillitatis was filled with basaltic flow about 3.7 billion years ago and Oceanus Procellarum (also Mare Serenitatis) about 3.2 to 3.4 billion years ago. "It would thus appear that the major mare surfaces were formed during the period 3.2 to 3.8 billion years ago and since then very little change has taken place in the major mare picture," they concluded.

Citing an earlier proposal by Wood of the Smithsonian Institution Astrophysical Observatory that anorthosite had constituted the primitive outer crust of the Moon, the New York group theorized that it took 400 to 500 million years for the anorthosite crust to cool. This would account for the fact that the Apennine sample was about 400 million years younger than the supposed age of the Moon (4.5 billion years). The anorthosite layer was bombarded later by incoming planetesimals to form the maria basins. But it was conceded that there was no evidence of a uniform layer of anorthosite over the entire Moon.

The phase of lunar activity for which scientists had the best record was the outpouring of lavas into the mare basins. The oldest basaltic rock was Rock No. 14,053, found at Fra Mauro. It was dated at 3.95 billion years old, about 200 million years older than the Tranquility Base basalts. The youngest basalts, found at Hadley-Apennines, were 3.15 to 3.36 billion years old.

These ages suggested a period of 800 million years during which volcanism had cooked lunar materials, although most of this activity had taken place in the major interval of 600 million years.

Earlier than 3.95 billion years ago, the Lunar Sample Analysis Planning Team noted, "the record grows dim, although we are confident that extensive igneous activity occurred in the lunar crust during the earliest period of its history."[13]

The anorthosite indicated an extensive melting epoch. At the Third Lunar Science Conference, experts were agreed that there had to be a very large amount of liquid to produce plagioclase-rich material. The most likely time this could have happened was at the beginning.

The high concentration of radioactive minerals in the western maria, comprising perhaps half of the total radioactive material on the surface, in Arnold's opinion,[14] required a separate explanation. Adler suggested that perhaps there had been a solid Moon at the beginning, and it partly melted inside, exuding a uranium-rich material that covered a large part of the surface.[15] The material was splashed about over wide areas by impacts.

Since the region of high radioactivity was centered around Imbrium, could the planetoid that created Imbrium have brought it to the Moon? This question was raised in several discussions at the conference. Arnold said he did not think so. He and his colleagues sought the explanation in lunar processes. Suppose, he said, that half the radioactivity of the Earth's surface was found to be in Australia and graded off toward

Indonesia and other regions. He put this situation to several colleagues and asked how they might explain it. None of them could, he said.

The results of gamma-ray study were confirmed by the alpha-particle detector on the Endeavour. The highest alpha-particle counts were recorded in the region of high-radioactive minerals especially near the Crater Aristarchus and in the vicinity of Schröter's Valley. These were the sites of frequently reported glows, mists, and fogs.

In their report on the alpha-particle experiment results, Paul Gorenstein and Paul Bjorkholm of American Science and Engineering, a research firm, said the increase in radon-emitted alpha particles must be associated with "optical events indicating the emissions of volatiles."[16]

In a later report, these researchers added that high particle counts were detected at the edges of the maria also, except for Mare Serenitatis.[17] The counts coincided with the "optical events" also seen at maria edges. They reported "a remarkable correlation" between the isotope polonium-210, a product of radon decay, and the edges of the maria. The experimenters hypothesized that radon is a minor component of the gases producing the "optical events"—about one part in 100,000. However, the results of the experiment established that at least one gas, radon, is being emitted from the edges of the maria, they asserted.

The X-ray and gamma-ray results were complementary. They not only mapped the broad areas of the mare basalt, KREEP, and anorthosite rocks, but they showed that the highest concentrations of radioactive minerals were in the Aristarchus area, in the high ground just west of the Apollo 15 landing site and south of the Crater Archimedes and in the region south of the Crater Fra Mauro.[18]

The Moon Is Trying to Tell Us Something

Supplementing the radioactive clocks on which the lunar dating game was based, was the cruder way of telling time on the Moon by counting craters. This method assumed that the rate of cratering diminished with time.

The radioactive clocks confirmed that the most densely cratered and smashed-up regions of the Moon were the oldest; the less densely pocked areas were younger. This could be seen quite readily in comparing the higher crater density in the highlands with the younger maria. Eugene M. Shoemaker, a geologist at the California Institute of Technology, related the density and distribution of craters to the dynamic process of accretion over time. The significance of this relationship was not confined to the Moon. It revealed something about the early history of the Solar System.

By the time of Apollo 15 a general notion of lunar evolution was taking shape. Indeed, more was becoming known about the early history of the Moon than about the early record of the Earth. The accretionary process that could be seen so vividly on the Moon indicated what must have happened in the evolution of the other terrestrial planets—Mercury, Venus, and Mars. It was a major aspect of the history of the inner planets. Shoemaker said that if one looked long enough and hard enough at the craters on the Moon, one became aware that "the Moon is trying to tell us something. It's telling us something about the accretion of the Earth."[19]

In order to decipher the message, Shoemaker first calculated changes in the cratering rate. In the interval between 3.65 and 3.85 billion years ago, the rate was at least 150,000 craters 1 kilometer in diameter over an area of a million square kilometers per billion years.[20] At 3.2 billion years ago and thereafter, it dropped to an average of 1,500 to 3,000 craters 1 kilometer in diameter per million square kilometers per billion years.

From these rates, Shoemaker concluded that the craters were made by two families of objects at different times. The earlier family consisted of planetesimals in solar orbit, swept up mainly by the Earth. Shoemaker estimated that if this swarm had a half-life of 45 million years (the time it took for half of them to be swept up), the entire early lunar surface would have been saturated with craters 1 kilometer or larger. No part of it older than 4 billion years would have escaped being smashed up.

The early planetesimal bombardment hypothesis would be brought to bear on the fission hypothesis, he said. For example, one of the main arguments against fission is that the angular momentum (spin) of the Earth–Moon system has been too low to have allowed the Moon to be spun off. If there was a primordial Earth–Moon planet that separated into the Earth and the Moon, it would have had to be spinning much faster than the Earth does now to cause the Moon piece to break off. But if that was the case, how was it slowed down to the present rate?

A fast-spinning Earth has been assumed by fissionists ever since George Darwin's time. He calculated that the primordial spin rate was between 3 and 5 hours—a rather short day! Rotation was slowed down over time by the friction of tides caused by the Moon and is still decreasing from this effect, Darwin maintained.*

* See Chapter 2.

Shoemaker invoked the early planetesimal swarm to slow down the rotation of both the Earth and the Moon by the braking effect of impact. The fission that created the Moon could have occurred when the Earth accreted only two-thirds of its present mass. If the final third accreted after the Moon was thrown off, then the Earth could have been "de-spun" to its present 24-hour period.

Furthermore, he added, the Moon could have lost substantial mass as well as angular momentum by heavy bombardment. It is possible that the impact velocities of the planetesimals were so high that more mass was splashed out of the Moon's gravitational field than the Moon gained. The higher gravity of the Earth would have prevented the splash-out from escaping. Shoemaker estimated that the amount of mass lost from the Moon by planetesimal bombardment could be equal to or even greater than its present mass.

So much for the first family of bombarding objects. Shoemaker suggested that the average rate of impact for the second family of objects hitting the moon since 3.2 billion years ago has remained fairly constant. The primary and secondary craters produced by these objects account for nearly all the craters formed since then, he said. "This family of objects can be shown with considerable confidence to be composed primarily of comets and fragments of comets."

It was possible to calculate what the tail end of the accretion process was like in terms of the cratering rate on the Moon, he told a news conference.[21] And that process could not have caused the large amounts of gravitational (impact) heating "that have been widely bandied about at this conference and its predecessors" to explain the chemical cooking of the rocks. "It wasn't fast enough to heat up the Moon to give a very early stage of melting, for example," Shoemaker argued. "I think the Moon is hot at the beginning, but it could not have been heated by the energy released by accretion."

The cratering rates suggested that accretion was spread out in time. Indeed, objects were falling on the Moon that had been stored somewhere in orbits for more than a billion years. The object creating the Crater Copernicus 850 million years ago must have roved about the Solar System since the planetesimals condensed at least 4.5 billion years ago. Shoemaker supposed that many of these later arrivals on the Moon were Mars-crossing asteroids, which had been stored in stable orbits for eons until perturbed out of them into a lunar collision course.

Edward Anders, a meteorite expert from the University of Chicago, joined the news conference discussion on this subject with a comment that there was evidence that meteorites accreted in less than 15 million years and possibly in only 2 million years. This would mean a considerable bunching up of objects in the orbits of the planets.

Anders told Shoemaker: "I think your model has the virtue that it is simple and you start with essentially a uniform distribution. On the other hand, there seems to be evidence that accretion was much faster than you would get on that basis." Perhaps, said Anders, the first stage of accretion was very rapid, involving maybe 80 to 90 percent of the mass in the orbit of the planets. Then the last 10 percent was thrown into longer-lived orbits and therefore lasted much longer. This would account for the decreasing cratering rates.

Shoemaker did not disagree with this. In fact, he said, the accretion model of the Earth he had in mind would "grow" its first 15 percent in a million years.

The length of time it took the planets to accrete could not be settled by exploring the Moon alone. It seemed to require a more comprehensive inspection of the Solar System—at least an Apollo-type exploration of Mars. The Moon, however, had furnished important clues.

One spring day at the University of Chicago some time later, Gerald Wasserburg alluded to the accretion problem. "Accretion could have taken from 6 days to 6 million years. Credence in the first citation has been devoutly expressed by earlier workers."[22] He was referring to the Book of Genesis.

Notes

1. Third Lunar Science Conference, News Conference, January 11, 1972.
2. Langsleth, Marcus G., et al., Heat Flow Experiment, NASA, Apollo 15 Preliminary Science Report, 1972.
3. Third Lunar Science Conference, News Conference, January 11, 1972.
4. Ibid.
5. Ibid.
6. Ibid.
7. Ibid.
8. Adler, I., et al., "Preliminary Results from the S-161 X-Ray Fluorescence Experiment," Third Lunar Science Conference, 1972.
9. Apollo 14 Preliminary Science Report, NASA. 1972. Preliminary Examination of Lunar Samples, Preliminary Examination Team.
10. Third Lunar Science Conference, News Conference, January 11, 1972.
11. Third Lunar Science Conference, Lunar Sample Analysis Planning Team, *Science*, Vol. 176, June 2, 1972.
12. Schaeffer, O. A., et al., "Age of a Lunar Anorthosite," *Science*, Vol. 175, January 28, 1972, and "Ages of Lunar Material from the Hadley Rille Area," . . . , Third Lunar Science Conference, 1972, abstracts.

13. Third Lunar Science Conference, News Conference, January 11, 1972.
14. Ibid.
15. Ibid.
16. Gorenstein, Paul, and Bjorkholm, Paul, "Detection of Radon Emanation from the Crater Aristarchus, . . . ," *Science*, Vol. 179, February 23, 1973.
17. Gorenstein, Paul, and Bjorkholm, Paul, "Detection of Radon Emission at the Edges of Lunar Maria," *Science*, Vol. 180, February 1, 1974.
18. Metzger, A. E., Trombka, J. I., Peterson, L. E., Reedy, R.C., and Arnold, J. R., "Lunar Surface Radioactivity," *Science*, Vol. 179, February 23, 1973.
19. Third Lunar Science Conference, News Conference, January 11, 1972.
20. Shoemaker, E. M., "Cratering History and Early Evolution of the Moon," Third Lunar Science Conference, 1972, abstracts.
21. Third Lunar Science Conference, News Conference, January 11, 1972.
22. Wasserburg, G., Lecture, University of Chicago, April 19, 1973.

11

Descartes

At the beginning of 1972 the main chemical provinces of the lunar surface had been roughly established. The low-lying maria, consisting of basalts enriched in iron and titanium and, in the west, in radioactive and rare earth elements, had been sampled by Apollo 11 and 12. An intermediate region between the maria basins and the bright highlands (terrae) on the south and east rim of the Imbrium basin had been sampled by Apollo 14 and 15. These missions found clues to the existence of an ancient, anorthositic crust, older than the mare basalts and richer in calcium and aluminum. The anorthosite appeared to be the main constituent of the highlands. At least, so the orbital surveys indicated.

Like Caesar's Gaul, the Apollo exploration of the Moon was divided into three parts: the maria, the circum-basin region of Imbrium, and the highlands. The maria, which appeared to be smoother and offer a safer harbor, were logically the sites of the early landfalls. As Apollo navigation technique improved and space agency officials gained more confidence in the performance of the lunar module, more difficult landfalls became acceptable. The landing of Apollo 15 between a steep mountain front and deep canyon demonstrated that men could land safely in the rugged highlands. This bright region constituted four-fifths of the area of the Moon's surface and appeared to be the most complex of the three Apollo areas of investigation.

Unlike the basaltic maria, the origin of the highland terrae could not be readily inferred from their appearance or general physiography.

Intense bombardment had obliterated obvious evidence of volcanism, if there had been any. But premission planners assumed that volcanism had filled the highland basins, as it had filled the maria. The hilly, hummocky regions were interpreted as "extrusive, igneous features, formed by viscous, silicic, igneous liquids."[1] Filled basins and hilly regions were common landscapes in the eastern, equatorial highlands on the front side of the Moon. High-resolution photos taken on Apollo 14 showed a central highland area 60 miles north of the Crater Descartes where both formations appeared to be accessible from a fairly smooth landing area. The U.S. Geological Survey had designated the basin fill as the Cayley formation and the hilly, hummocky areas as the Descartes formation.

There were two, young, bright-rayed craters in the area, one north and the other south of the landing site. Dome-shaped hills rose in the northeast and southeast, representing a projection of the Descartes formation into the Cayley plains. The hill to the north was dubbed Smoky Mountain. The one to the south was Stone Mountain.

The young, deep craters had been punched below the regolith and it was likely that their ejecta consisted of ancient anorthositic crust as well as igneous rock assumed to overlay the primitive crust.

This area of the Cayley-Descartes formations became the landing target of Apollo 16 and the search for evidence of highland volcanic and plutonic processes its main objective.

The predominance of anorthosite in the highlands as indicated by orbital surveys was supported by analysis of a highland grab sample picked up by the Russian automatic lander Luna 20. The robot scoop landed in the highlands between Mare Fecunditatis and Mare Crisum in February 1972 and brought back 2 kilograms of soil and rock chips—all by radio control. The Luna 20 sample had not indicated the presence of volcanic rock, but 3 days of wide-ranging investigation by two astronauts were expected to give a more complete picture of the highlands.

The region of the Descartes-Cayley formations was called Descartes for the crater that memorialized the French philosopher-mathematician, René Descartes (1596–1650). Topographically, the region at 9 degrees south latitude and 15 degrees, 30 minutes east longitude was one of the highest on lunar near side, with an elevation of 7,000 feet above the mean level of the lowland maria.

From orbit the Cayley formation looked like an undulating plain covered by lavalike material, which was thought to have flooded the highland craters somewhat earlier than the flooding of the mare basins. The two craters, each about a half mile in diameter, penetrated deeply into the crust. Large blocks could be seen on their rims and were believed to be samples of the basin-filling material that the explorers could readily collect for analysis. Because they were north and south of the landing site, the craters were dubbed North Ray and South Ray.

The hilly, furrowed area of the Descartes formation appeared to be similar in age to but of different composition from the Cayley formation lava flows. Beneath both igneous layers, mission scientists expected, the explorers would find pieces of the ancient crust of the Moon.

The basin-filling material supposedly represented by the Cayley formation was fairly widespread in the highlands, covering about 7 percent on near side. The Descartes formation seemed to be typical of about $4\frac{1}{2}$ percent of the near side highlands. The rocks there were thought to be a type of basalt quite different, chemically, from the mare basalts.

There were high hopes for Apollo 16, which was to bring a second heat flow experiment to the Moon; conduct a second active seismic experiment, with the thumper and grenade launcher; install a fourth operating passive seismometer to enlarge the network; bring both stationary and portable magnetometers; sample the top 500 microns (0.5 millimeter) of the surface to detect solar wind and cosmic radiation effects; make the first analysis of cosmic rays from the lunar surface; take the first far ultraviolet light photos of the Earth and of the galaxy from the surface with a far ultraviolet camera and spectrograph provided by the Naval Research Laboratory; and continue the orbital experiments performed on Apollo 15.

In the opinion of William R. Muehlberger, the mission's field geology principal investigator, Apollo 16 and 17 highland missions were expected to return more information than all four previous landing missions combined.[2] For this reason, the data they were expected to return represented half of those of the entire Apollo lunar investigation, he said.

The Layering Question

Apollo 16 would seek highland volcanic material and Apollo 17 would look for the oldest and the youngest rocks on the Moon. Reconaissance photos of the Taurus-Littrow region of the highlands where Apollo 17 was targeted indicated recent volcanism there as well as a primitive crust. Evidence of volcanic activity younger than 3.1 billion years, the age of the youngest rocks found so far, would be a major discovery. However, all the evidence showed that the lunar heat engine shut down three eons ago. Except for the possibility of small, residual pools of lava deep underground, the Moon was volcanically inert.

The highland missions might shed additional light on the apparent layered structure of the mountains and rilles, so visible to Scott and

Irwin in the Hadley-Apennines. At the Third Lunar Science Conference there was rising interest in this phenomenon. How did it take place?

On Earth layers of rock, usually tilted at various angles, are the products of sedimentary deposits in the sea. They are being formed now in such places as the deltas of the Nile and Mississippi rivers. Eventually, the compacted layers are uplifted as mountains are formed. Then a highway department makes a cut through the mountains to build a turnpike and the layers are sharply exposed in cross-section, often banded in different colors.

Layering on the waterless Moon represented a different process. It might be the result of intermittent lava flows. In the Apennines, successive blankets of ejecta that were heaved out of craters and piled up in the dunelike Apennine Mountain arc by the Imbrium impact might account for the layering Scott and Irwin saw and photographed.

The Earth showed one mountain-building mechanism; the Moon, another. Mars might yet reveal a third. In the early months of 1972 it was becoming evident that answers to planetary evolution would require the investigation of a third planet, Mars. And the importance of Mars as the next port of call in the exploration of the Solar System was becoming more obvious as Apollo rolled on. For the exploration of Mars, however, manned missions were highly speculative, not only because of cost but also because of the unresolved question of whether human beings could endure the journey. The post-Apollo Skylab program calling for long-duration missions in the Earth's orbit was designed to answer that question. In the meantime, both the United States and Soviet Union were extending tentative toes to test the waters around Mars. Robot landers were planned by the Russians in 1974 and the Americans in 1975, in addition to orbital photo reconaissance vehicles.

Orion and Casper

Still seeking the primeval crust, the space agency prepared for the flight of Apollo 16 to the Descartes highlands near the center of the Moon. It was the most ambitious mission so far.

The launch had been scheduled for March 17, 1972. Early in January, NASA announced a month's delay in order to strengthen the Moon suits, which had been made more flexible, recheck the lunar module descent batteries, and rework a docking ring jettison device to ensure a clean separation between the lunar and command modules.

Apollo 16 was launched from the Kennedy Space Center at 12:54 P.M., EST, April 16, 1972. Commander of the mission was Navy Captain John W. Young, 41, a test pilot who joined the Corps of Astronauts in 1962 and flew the first three-orbit Gemini flight with Virgil I. (Gus) Grissom March 23, 1965, in the good ship Molly Brown. Now, just 7 years later, he was en route to the Moon, a prospect that was only a faint hope in the early days of the two-seater Gemini spacecraft.

Young helped make space history a second time as command pilot on the 3-day flight of Gemini 10 with Michael Collins in July 1966. On this flight, the Gemini docked with an Agena rocket that was ignited and propelled the astronauts to a record altitude of 475 miles. It was the first time a large rocket had been used in space to move a manned spacecraft, and there was speculation at the time that this kind of ma-neuver could be used to send a crew around the Moon.

On his third space mission, Young served as command module pilot on Apollo 10, which flew to lunar orbit in May 1969, the final practice flight before the landing of Apollo 11 the following July. On this flight, the commander, Thomas P. Stafford, and the lunar module pilot, Eu-gene A. Cernan, descended in the lunar module to within 8 nautical miles of the lunar surface in a test of the LM descent and ascent systems. They located and photographed the Apollo 11 landing site and also experienced the gravitational effect of the lunar mascons in ac-celerating an orbiting vehicle. This effect subsequently caused the Apollo 11 overshoot of nearly 4 miles.

Born in San Francisco, Young entered the Navy upon graduation in 1952 with a Bachelor of Science degree in Aeronautical Engineering from the Georgia Institute of Technology. He subsequently received hon-orary degrees of Doctor of Laws from Western State University in 1969 and Doctor of Applied Science from Florida Technological University in 1970. Young was married to Susy Feldman of St. Louis and had two children by a previous marriage, Sandy, 14, and John, 13. He was a Fellow of the American Astronautical Society, and an Associate Fellow of the Society of Experimental Test Pilots. In 1962 he set two world altitude climb records in the F4B Phantom to 3,000 and 25,000 meters.

Young's partner on the Moon was Air Force Lieutenant Colonel Charles M. Duke, Jr., 36, born in Charlotte, North Carolina. Duke was graduated from the U.S. Naval Academy with a Bachelor of Science degree in Naval Sciences and subsequently joined the Air Force. He received a Master of Science degree in Aeronautics and Astronautics from the Massachusetts Institute of Technology. He and his wife, the former Dorothy Claiborne, had two children, Charles M., 7, and Thomas C., 4. Duke was also a member of the Society of Experimental Test Pilots. He was an instructor at the Air Force Aerospace Research Pilots' School in 1966 when he was selected as an astronaut.

The command module pilot on this mission was Navy Lieutenant

Commander Thomas K. Mattingly, II, 36, who had been scrubbed from the flight of Apollo 13 just 72 hours before its launch because of exposure to German measles. At last, Mattingly was getting his chance at the Moon.

The command module had the call sign Casper and the lunar module, the call sign Orion.

During the coast to the Moon, two problems appeared. As they left the Earth behind and photographed it with the ultraviolet light camera, Young, Mattingly, and Duke noticed they were entering a region of space swarming with particles. These were reminiscent of the "fire flies" John Glenn had seen during his three-orbit flight in 1962. The particles swarming around the Apollo command module, like the particles Glenn saw, came from the space vehicle itself. They were flakes of paint from the lunar module.

Analyses by the LM manufacturer, the Grumman Aircraft Engineering Corporation, and by NASA indicated that the particles "snowing around" Casper were pieces of a painted aluminum skin, applied to certain parts of the LM to prevent it from overheating in the Sun.

No sooner had this problem been disposed of than another appeared. At 2:04 P.M., April 18, when Casper and Orion were 139,000 miles out from Earth, Mattingly reported that the guidance and navigation system had gone awry. It gave no altitude indication and the gimbal platform had locked. At Mission Control in Houston, a warning light appeared on the Guidance Officer's console.

With instructions from Houston, Mattingly freed the platform that immediately became inertial. His next task was to realign it. In order to do this he had to take star sightings, but Mattingly informed Houston that the "snowflakes" that were traveling right along with the spaceship interfered with star sightings. He succeeded in aligning the platform mainly by sighting on the Sun. When he completed this task, the inertial measuring system—without which even lunar orbit could not be attempted—remained stable and the tension at Houston relaxed.

Later Mattingly was able to make fine adjustments to the alignment when he could get a clearer view of the stars. No one knew what had happened to throw the inertial guidance system into a lock. It was probably one of those mysterious, electrical "transients" called a "glitch," that no one really understands but that create temporary malfunctions in electronic equipment and then disappear without a trace. As a safeguard, the guidance computer was programmed not to respond if a subsequent "glitch" occurred during a maneuver.

Casper and Orion sailed into lunar orbit at 2:22 P.M., CST, April 19. Another problem appeared. This time the S-band steerable antenna on

Orion jammed and could not be aligned with the Earth. Communications became difficult. The omnidirectional antenna had to be used for the remainder of lunar operations until the ground S-band antenna was set up on the Rover after landing.

On the afternoon of April 20, Orion separated from Casper and all was in readiness for Young and Duke to descend to Descartes when another new problem cropped up in the Apollo spaceship. Mattingly reported a bad circuit in the backup steering system controlling the ship's main rocket engine. The prime system was in working order, but the backup system, to be used if the prime system failed, was causing the engine bell to oscillate sideways.

Mattingly had noticed this when he attempted to circularize Casper's orbit on the thirteenth revolution of the Moon. Although the prime steering (thrust vector) control system was in good order, mission rules required that the backup system had to be working perfectly as well as a condition for a landing. Otherwise, the lunar module's descent propulsion system would have to be used as a backup for the boost out of the lunar orbit and the return to Earth. If the backup system was malfunctioning, the lunar module could not be permitted to land.

The reasoning was straightforward. If only the prime steering system was available and it failed, the crew would be marooned; without the backup, there was no way of returning home.

A crisis atmosphere thickened in the Manned Spacecraft Center. Young and Duke were instructed to go around the Moon again and stay close to Mattingly in case it was necessary for Orion and Casper to redock and return to Earth. For a while, it appeared that another Apollo 13 was in the making.

The Houston flight directorate advised the crew that the steering problem was being analyzed at Houston and by the manufacturer, North American-Rockwell at Downey, California. It had been decided to allow five revolutions of the Moon (about $7\frac{1}{2}$ hours) to see if the problem could be solved before the landing was scrubbed.

In tests Mattingly ran in consultation with Houston, the problem was pinpointed to a defective yaw gimbal actuator. This device was supposed to swing the engine bell right or left when signaled by the pilot or the computer. However, when activated, it caused the engine to oscillate back and forth, giving rise to fear the oscillations could shake the spacecraft apart. This problem appeared only in the backup system, not in the prime guidance system. It was probable the backup system would never be used for the return to Earth, but rules forbade taking that gamble.

At Downey and at Houston, engineers were trying to determine whether these oscillations would damage the vehicle in the event this system had to be used to get out of lunar oribt and head back to Earth. They diagnosed the problem as an open circuit in a feedback system that controlled the engine gimbal (swing).

After several hours, it was decided at Downey and Houston that the yaw oscillations would not affect the attitude of the ship and would not damage it structurally if this backup guidance system had to be used instead of the prime one. This was good enough for the flight director-ate. At 5:44 P.M., CST, Christopher Columbus Kraft, the Manned Spacecraft Center director, announced without preamble the decision to go for landing on the sixteenth revolution.

The decision was passed up to Orion as it and Casper came around the eastern limb of the Moon on their fifteenth revolution. Young and Duke were delighted. It had been agreed that, because they would be landing 6 hours later than scheduled, they would get a night's rest and commence surface operations on the morning of April 21.

They landed Orion at 8:23 P.M. on the Descartes formation, 230 meters (747.5 feet) northwest of the planned landing site. They were at 8 degrees, 59 minutes, and 34 seconds south latitude and 15 degrees, 30 minutes, and 47 seconds east longitude.

The Orange Juice Menace

After checking Orion's systems, Young and Duke had supper of dehydratable cream of tomato soup, rye bread, a tuna spread, an apple food bar, a chocolate bar, and an orange-grapefruit drink. It was the Day 5 Meal B. All the calories, vitamins, minerals, and roughage had been counted carefully. Somehow, the tomato soup usually came out cold, though.

In order to boost muscle tone, the space agency had equipped the Moon suits with an orange juice container in the neck piece. It allowed the wearer to sip orange juice through a tube, not only as a refresher but also to increase dietary potassium with which the orange juice was richly salted. Loss of potassium during zero gravity flight was equated with loss of muscle mass in crews. With extra potassium in the orange juice, the NASA medical staff believed muscle loss could be reduced and muscle tone preserved.

The experiment created several problems, however. Preparing for the first EVA, Duke spilled some of the juice refresher and got it into his hair as well as the inside of his helmet. The headpiece had to be cleaned thoroughly to prevent the juice from interacting with the visor defogging compound.

"I wouldn't give you two cents for that orange juice as a hair tonic," Duke told Anthony England, a geologist astronaut who was the explorers' capsule communicator at Houston. The juice matted his hair down completely, Duke complained.

Duke described the ground around the LM as rolling, hardly flat anywhere, and liberally sprinkled with rocks. It was hummocky with 4- to 5-meter ridges. In the morning sunlight, the rocks looked black and white. About 50 meters beyond the LM, there was a ridge that obscured the landscape beyond.

"It's neat to have a gravity field around to set stuff on," Duke commented. "I feel like a little kid on Christmas eve. Looking toward the horizon, there's a very hilly, subdued region disappearing from 12 o'clock to 11. It's rolling, pocked with white craters. At 1 o'clock, I can see 1 kilometer or so—more rolling terrain of similar albedo (reflectivity), a light gray with fresh, white craters. At 3 o'clock, a nearby ridge blocks out North Ray Crater and Stone Mountain. In front of the LM, at 50 to 100 meters, is a low ridge and beyond it a depression, then another ridge which goes into Spook Crater (which Duke thought was on his horizon at 12 o'clock)."

The boulders they could see out the window were not impressive, only 2 to 3 meters wide. As they were descending, they saw a distinct ray pattern across the landing site. It was very marked from an altitude of 5000 feet. Stone Mountain appeared terraced and displayed the same kind of lineations that Scott and Irwin had seen on the Apennines.

Young and Duke went to sleep about 11:30 P.M., while Mattingly sailed by overhead in Casper in an orbit of 66.8 by 53 nautical miles. During the early hours of Friday morning, April 21, music filtered down between the spheres, seeping out of the background radio noise of the cosmos. Mattingly was playing *Symphonie Fantastique* by Hector Berlioz on his tape recorder. Now and then, he discussed the scenes he was photographing with Stuart Roosa, the capsule communicator for Casper. He took a strip of photos as he sailed over the eastern rim of the Crater Ptolemaeus, one of the grand sights on the Moon.

As he was approaching the sunrise horizon, preparing to photograph the zodiacal light, Mattingly saw a bright flash south of his ground track and several degrees below the horizon. It was not repeated. And the nature of it was not determined.

The First EVA

The first day on the Moon began with high expectations and verbal flourishes. Young climbed down to the surface at 10:50 A.M. and de-

APOLLO **16** TRAVERSES

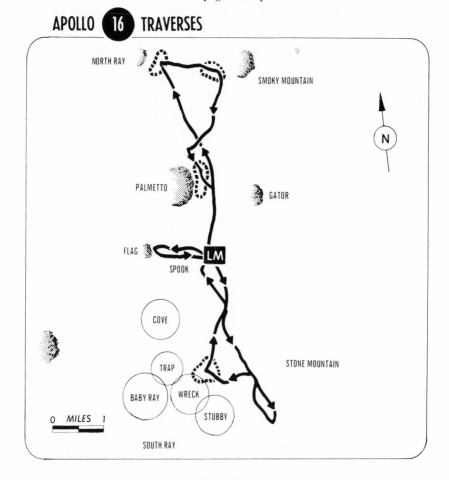

NORTH RAY

SMOKY MOUNTAIN

N

PALMETTO

GATOR

FLAG
SPOOK
LM

COVE

TRAP

STONE MOUNTAIN

BABY RAY   WRECK

STUBBY

0   MILES   1

SOUTH RAY

livered an impromptu oration: "Hello, mysterious and unknown Descartes. Apollo 16 is going to change your image."

Little did he know then how right he was. The whole preflight conception of Descartes was going to change during the 3-day sortie.

"Oh, look at those beautiful rocks."

Duke came down the ladder and said: "Fantastic . . . this first foot on the lunar surface is super, Tony. Okay, we're making little footprints here about one-half inch deep—not kicking up very much. Good lord! Look at that hole we almost landed in!"

At Houston's insistence, they examined Orion's steerable antenna, which would not steer, but could not find anything obviously wrong with it outside as well as inside. Then they pulled out Rover II, the second automobile on the Moon.

"Me and Charlie . . . we just picked up the vehicle," said Young, strong-man style.

"You don't know your own strength," said Tony England at Houston.

At first, the rear steering wouldn't work, but that was par for the flight. They set up the lunar communications relay system and mounted the television camera on the Rover, pointing the umbrella-style S-band antenna at Earth, which was directly overhead. The first pictures to reach Houston showed Young raking rocks near the Rover. They then set up the United States flag and saluted it in front of the TV camera. Young snapped his arm up so vigorously for the salute, the action lifted him off the ground. As they started to unload and set up the ALSEP experiments, Houston called up and said: "This looks like a good time for some good news here. The House passed the space budget, 277 to 60, which includes the votes for the shuttle."

"Beautiful, beautiful," said Young.

"Wonderful, wonderful," said Duke. "Tony, again, I'll say it with salute. I'm proud to be an American."

"So am I," said Tony England.

"Man, I'll say it, too," said Young. "The country needs the shuttle mighty bad. You'll see."

In the shadow of Orion, John Young set up the ultraviolet camera-spectrograph on a tripod. The 75-millimeter electronographic Schmidt camera, with its magazine of 35-millimeter film, constituted the first astronomical photographic experiment on the Moon. With assistance from Mission Control, Young aligned it to photograph hydrogen around the Earth and sources far beyond, in various regions of the galaxy.

Although physicists have speculated that hydrogen exists throughout intergalactic space, and clusters of gas may be detected between the great galaxies, observations by satellites in low Earth orbit were thought to have been impaired by the masking effect of the Earth's corona of gas. The Moon provided an ideal platform for ultraviolet astronomy.

Inspection of the landing site for a level place to set up the ALSEP instruments convinced the astronauts they had been lucky indeed to have found such a level place to land. Even the craters have craters, Duke remarked. They found a place to set the ALSEP 300 feet south of the LM.

Using an improved surface drill, Charles Duke had no difficulty drilling the first 10-foot hole for the heat flow experiment that, at long last, was about to be installed as per specification. The fiberglass rod with sensors and probes went into the hole beautifully. Duke then laid out the cable and prepared to drill the second hole when John Young, bounding along, got his foot caught in the cable and yanked it off the heat probe rod. The cable broke off exactly at the connector.

YOUNG: Well, I guess I can forget the rest of the heat flow.
DUKE: Oh, rats!
YOUNG: Sorry, Charlie. Goddam! You know it!
DUKE: I know you are. A bunch of spaghetti over there.
CAPCOM: Boy, we can sure see that on the TV. It looks like a mess.
DUKE: Well, tell Mark [Marcus Langseth, the principal investigator]
   we're sorry. There's no way we can recover from that, Tony?
CAPCOM: Well, we're working on it.

Duke used the drill again to get a deep core sample and applied a jack to extract the core tube. The jack was added to the equipment as a result of the problems Scott and Irwin had experienced extracting the drill core on Apollo 15 at Hadley-Apennines.

After laying out the passive seismometer, the stationary magnetometer, and setting up the other experiments, they deployed the geophones for the active seismic experiment. Young then used the thumper to generate signals, leaving Duke to examine the torn heat flow cable.

Duke picked up a solid piece of glass. It was spherical. He held it up in front of the TV camera. Young, meanwhile, set up the mortar package on the active seismic experiment. It would be fired after the explorers departed.

Suddenly, Duke started coughing and a moment later explained that the "in-helmet orange juice went down the wrong way."

Early Friday afternoon, April 21, Young and Duke climbed into the Rover and set out across the plain for Flag Crater, about 1.3 kilometers to the west. Duke told Tony England that South Ray Crater was visible. It had "a tremendous amount of blocks in it, with some black streaks."

Young, at the controls, steered through a boulder field and navigated around craters. The ground was much rougher than it had appeared in the photographs, which had showed the Cayley Plain as relatively smooth and broadly undulating. Their first destination was Station 1 near the rim of Plum Crater, a depression 30 meters in diameter on the rim of the 300-meter Flag Crater. En route, according to the map, they should pass Spook Crater, a large one where they would make a second stop to collect samples.

As their predecessors had found it, it was not easy to tell where you were in hilly, cratered regions. They thought they passed Spook Crater on their left, but weren't sure. Driving on, they looked for another landmark, Buster Crater. Duke thought he saw it.

"There's a crater over there," he said. "A big one."

"Boy, that's a biggie," Young agreed.

They were sure it had to be Buster Crater, but England advised them that Buster was only 40 meters in diameter. The crater they were look-ing at was larger. It had boulders in the bottom, rocks some 5 meters in size.

Westward, they approached the rim of a crater they were sure was Flag. There, they hoped to get samples of the Cayley formation. Flag Crater was on a ridge that sloped away to the southwest, toward South Ray Crater. On the north side of Flag, Duke said, there was a crater on the inner rim, presumably Plum Crater. They had arrived.

Looking down into Flag Crater, Duke reported that the sides were so steep, he could not see the bottom from the rim. This was to be a common experience. In Plum Crater he described a partly buried boulder about 1 meter across, but no bedrock.

Both explorers began to collect samples. Duke described a breccia. It consisted of a white clast with greenish-looking crystals about a milli-meter in diameter, in a black matrix. This kind of breccia was to become increasingly common as the predominant rock in the region during their 3-day stay. John Young picked up a breccia with white clasts and lineations. Duke then found a breccia with a white matrix, coated with glass on one side. They took a soil sample under the top regolith layer and found it was white underneath. Then they knocked off a piece from a boulder with the geology hammer. The piece contained a greenish clast and a whitish clast in a grayish matrix. The clasts were a millimeter in diameter and made up perhaps 5 percent of the rock. The largest crystal was 5 millimeters across.

Now and then, the astronauts would use a tongs to pick up rocks, but often they were able to lean over, especially on an upslope, and pick them up. Sometimes, they leaned on the shovel. One football-sized rock weighed about 20 pounds (Earth weight).

Turning back to the east, they drove to a crater they thought was Spook. While Young took a reading on the portable magnetometer, Duke shot a picture of South Ray Crater in the distance with the 500-millimeter telephoto lens of the Hasselblad camera.

As they approached the landing site area, Young tested the Rover at maximum acceleration, which turned out to be 17 kilometers an hour over this rough surface. This was "Grand Prix" driving to see how the Rover performed at speed and on sharp turns. When they parked near the LM, Young turned on the TV and Houston swiveled the lens about to observe the sight. At one point, the camera picked up the Earth.

ENGLAND: Hey, fellows, we're able to see the Earth with your Big Eye there.

DUKE: Pretty sight, isn't it.

ENGLAND: .Sure is. (Seeing the Earth he was sitting on from 238,000 miles away.) Yes, it's weird.

It was getting dark in Houston as Young and Duke at Descartes finished their first day on the Moon and began moving back into the cabin of Orion. Their EVA had lasted 7 hours and 11 minutes.

Young and Duke were tired and dirty—so dirty, Young said, he could hardly believe it. Their fingers were tired and sore from the pressure in the gloves—a problem that was becoming conspicuous now that the EVAs were getting longer.

Their conversation came over the public address system at Houston as that of two, weary men who were just beginning to relax after a hard day's toil.

YOUNG: I'm tired, too. I tell you.

DUKE: Okay.

YOUNG (doffing his helmet): Hey, that Moon dust . . . don't taste half bad.

DUKE: It that what it is?

YOUNG: First thing I want is a drink of water. I'm going to take a break and get me a drink of water.

DUKE: I finished mine long ago.

YOUNG: I can't believe I am so dirty.

After getting out of their suits and cleaning up the LM cabin, Young and Duke were to have their evening meal and begin an 8-hour rest period. Meanwhile, they talked back and forth with Houston. Duke told Tony England: "I tell you, Houston, my general impression of this thing is I'm a lot more surprised how really beat up this place is. It must be the oldest stuff around because it's just craters on top of craters on top of craters. And there are some really big, old, subdued craters that we don't even have mapped on our photo map."

Shortly after 6 P.M. Friday, Edgar Mitchell, who had flown Apollo 14 as lunar module pilot with Alan Shepard, came on duty as capsule communicator and Duke called down: "Hey, Ed, this is really a spectacular place. Now I know why you were so excited at Fra Mauro."

Debriefing

Because the directional high-gain antenna on Orion was out of commission as a result of some still undiagnosed malfunction, the crew had to rely on the LM's omnidirectional antenna while in the cabin. It was less efficient than the Rover high-gain antenna outside. Inside the LM, their communications were garbled by static. Communication from the LM would improve later in the evening, however, when the 210-foot dish at Goldstone, California, rose into radio view of Descartes on the Moon. The Apollo 16 crew and Mission Control officials and scientists awaited this alignment for an evening debriefing after that first EVA.

When the communications strengthened, Young remarked that the rocks they had found in the Cayley formation were all breccia, having a gray matrix and dark clasts. At Buster Crater, the rocks were so friable they crumbled in his hands. There were no volcanic rocks.

Asked whether they had sampled representative types, Duke replied he could not be sure. Near the LM, he said, there were some rocks with a pinkish-hued clast in them. Young broke open a boulder that revealed a whitish matrix within, containing clasts. But the predominant rock was just breccia.

These findings were to characterize the results of the whole mission, much to the surprise of the geologists who expected evidence of volcanism to explain the land forms. But the results of the first traverse were not at all convincing that brecciated rocks, composed largely of anorthosite fragments, would be all there was. There had to be igneous rocks there, too.

Did they see any outcrops? No, Young said. There were no outcrops and no bedrock. But he added they couldn't get far enough out on the rims to see the bottoms of Flag or of Spook Craters.

Duke was still enthusiastic after the long, hard day. He said he was looking forward to the next traverse. "The day went so fast today, the first thing I knew I didn't have a chance to eat or get a cup of coffee or anything. Doggone exciting!"

"It was pretty interesting," said Young. "I think we can do a little better on the driving across Sun tomorrow."

They would head south then.

"You made good time coming back," said Houston.

"Yeah. Follow your tracks. That's the only way to fly," said Young.

After the short debriefing, Young and Duke, thinking the microphones were off, began to converse in terms more visceral and explicit than they would have used if they thought they were still being heard over the Center's public address system. Young's microphone was still on.

He was complaining of flatulence and passing gas. He had acid stomach. "I mean I haven't eaten this much citrus fruit in 20 years. . . . I ain't never eating any more ... . and if they offer to serve me potassium with my breakfast I'm going to throw up. I like an occasional orange, I really do, but I'll be damned if I'm going to be buried in oranges."

As the conversation progressed, Houston, ever alert to protecting the astronauts' Rover Boy image, sent up a call.

CAPCOM: Orion, Houston.

YOUNG: Yes, sir?

CAPCOM: You're on your way to having a hot mike.

YOUNG: Oh! How long have we had that?

CAPCOM: It's been on through the debriefing.

YOUNG: How could we be on a hot mike with normal voice?

CAPCOM: John, would you exercise your push-talk button? It may be stuck.

YOUNG: Yeah, Houston.

CAPCOM: John, it doesn't seem to be a hot mike now. Evidently, you got it off. Advise a 7-hour EVA for the second day and 5 hours for the third.

YOUNG: I want to apologize for the hot mike. It's a terrible thing to have a hot mike here, sometimes.

CAPCOM: Well, you guys have done commendably well considering the fact you didn't know you were on it.

They settled down for their rest period. Duke took a seconal to help him sleep. All was quiet, then, at Descartes.

On the dark side of the Moon, Earthshine provided good illumination of the surface, which resembled a snowscape on Earth under a full Moon. But the lunar surface was even brighter in Earthshine. In sunlight, it was very bright and painful to look at through the sextant and the telescope. Mattingly said he could feel the heat coming up from the sunlit surface through the optical instruments.

The Second EVA

Saturday, April 22, was another big day at Descartes. After breakfast Young and Duke were to drive to South Ray Crater. There they expected to find the volcanics they sought.

The breakfast they tackled consisted of peaches, beef steak, bacon squares, spiced fruit cereal, something called Instant Breakfast, orange-grapefruit beverage, and a cherry food bar. Duke was afraid to eat it all. If he did, he said, he was sure he couldn't get his Moon suit on.

Back at CapCom, Tony England called up to advise that Mattingly could see benches or terraces in Stone Mountain, one of the goals of their Saturday traverse. Communications were shifting around the world as the big Australian antenna at Honeysuckle Creek came into view of

Descartes. For a few moments, Apollo 16 lost contact with Houston and talked to the station in Australia.

YOUNG: Okay. You guys are nice to talk to. We don't care about Houston.

HONEYSUCKLE: Well, thanks very much. Certainly appreciate it. It has been a pleasure working with you on this mission.

YOUNG: Roger. We'd sure like to come down there and see you folks.

HONEYSUCKLE: Well, you've got a permanent invite . . . any time you like . . . we'll keep the beer cool for you.

DUKE: Best idea I've heard all day.

Houston came on the line and England was saying that Station 8 on the traverse to Stone Mountain and South Ray Crater was a key station. They might get some of the materials Mattingly saw in South Ray Crater—big blocks of it. Duke asked what day it was and England informed him it was Saturday morning.

When the explorers once more descended to the ground, they were standing in hot sunshine. The temperature had risen to 130°F (but it was 150°F below zero in the shade). Looking south, Duke reported that the lineations in Stone Mountain now looked more distinct.

Stone Mountain was a dome rising 540 meters (1755 feet) above the Cayley Plain, a portion of the Descartes highland structure.

It was 11:25 A.M. before the explorers were ready to drive off to the south. As they bumped along, Duke reported that the boulder and cobblestone population was increasing. The largest blocks were about a meter in size. The regolith appeared to be loosely compacted and the Rover tracks were quite definite. Dust-covered rocks were mostly rounded and angular rocks seemed to be free of dust, Duke observed. It was evident the dust-covered rocks were older. Duke served more or less as the chronicler of the traverses. He described the regolith "as a freshly plowed field that's been rained on."

Presently, Duke reported, he could see the rim of South Ray Crater as "spectacularly white." It stood out above the surrounding terrain, he said.

They reached a broad area of high ground called Survey Ridge on their maps. Part of it was strewn with rubble, a ray of material from South Ray Crater. Driving through this boulder field was "terrible" and Young got out of it as quickly as he could.

Most of the rocks were 25 to 35 centimeters in size, but some were as large as 2 meters. Some of this debris came from a smaller crater than South Ray called Baby Ray. The stark, bright rays from Baby Ray were lying atop those from South Ray, showing that Baby Ray was younger.

As the explorers jolted southward the block population increased, and it became evident that it would be too thick to allow them to drive to South Ray's rim. They headed for Stone Mountain, their prime objective. At close range the lineations they had seen from afar were not visible. On the flank of the big dome, the rocks looked the same as on the Cayley Plains.

DUKE: That Stone Mountain looked like it was right on top of us and we've come 2.6 kilometers and it still looks just as far away. Boy, it's spectacular looking out to the west. That poop about being able to see the LM all the way on traverse two I think is going to be bum dope. We've come down some big swales.

YOUNG: They call them swales in your part of the world, Charlie. They call them mountains in mine.

Although neither of them realized it, the explorers were going up the mountain. The Rover took the grade so readily that they could hardly believe the instruments reporting a 10-degree slope.

ENGLAND: Hey, fellows, Ken was just flying over and he saw a flash on the side of Descartes. He probably got a glint off you.

DUKE: Yeah. That's us. Men of miracles. Looking back east, we see ravines. We see the rim of North Ray—some pretty good blocks on it. . . . We cannot see into North Ray . . . it's above acquisition, but we can see the whole lunar module.

They were high on the mountain now. Young said it was time to park. As they prepared to collect samples, Houston announced a decision not to attempt to repair the heat flow cable.

Duke took a series of 500-millimeter-lens photos of South Ray Crater. He reported that at their altitude on Stone Mountain he could see into the rim on the inner wall. He also photographed North Ray and "Old Orion."

DUKE: Wow! What a place! What a view! Isn't it, John?

YOUNG: Absolutely. Unreal.

ENGLAND: We're darn near speechless down here. [He was referring to the view on the control room television screen.]

DUKE: Can you guys see how really spectacular the view is?

ENGLAND: We sure can.

The camera's "big eye" swiveled about magically as Houston, controlling it by radio, pointed it at the dusty explorers.

"We're looking at you," said England.

"Look up-slope, Tony. Look up-slope," urged Duke. "And you'll see all this rock field we're in here. The block population is 60 to 70 percent. Most of the rocks are white clasts with a little glass coating on some."

"Charlie, you're on the Big Eye," said England. The men on Earth were watching him operate the penetrometer staff, testing the density of regolith on the mountain side. The lens moved to Young, who was digging a trench with the shovel. No layering was visible. He scooped a sample from the deepest part of the trench and poured it into Bag 399. Then they drove a double core into the regolith, extracted it and capped it.

This stop was designated as Station 4 on the map. From there, they headed downhill to Station 5. The drive was easy "as long as the brakes hold," Young noted. He had the power off and they were coasting down the mountain at 10 kilometers an hour.

"We're just falling down our own tracks," said Young.

"Probably a good idea you couldn't see how steep it was going up," said England.

"Darn right it was," said Young.

At Stop 5, they parked on the rim of a 15-meter crater. They alighted stiffly and began raking up samples, looking for rounded rocks. Houston asked for a documented sample of a glass-covered rock. This meant they must photograph the rock in place, with the gnomon in the picture for size, color, and local vertical reference.*

Duke found that most of the rocks he picked up crumbled as soon as he pinched them. There was no hint of bedrock. He and Young picked up several white rocks with bluish crystals until they found a glass-covered rock. They photographed it.

"One thing about being on a 20-degree slope—you can get down on your knees," Young said. He picked one rock after another, only to find it as friable as a clod.

"We're going to have to press on after this sample," warned England, watching the clock in Mission Control.

"Okay," said Young. "Twenty minutes to get back to the Rover."

They had walked far afield from their trusty vehicle.

---

* The gnomon consisted of a weighted rod attached to a tripod so that the rod would hang vertically when the tripod is on the ground. The staff had a gray scale and a color scale of blue, orange, and green. Adapted from an ancient device to measure the Sun's elevation, the gnomon was used on the landing missions for orientation of sample photos.

When they got back to the Moon car, England said, the plan was for Young to take a portable magnetometer measurement and for Duke to collect some samples around the rim of the crater on which the Rover was parked.

At the ALSEP site, the emplaced magnetometer was indicating a magnetic field of 130 gammas, England told them. It was the highest magnetic measurement yet made on the Moon. The field at Station 5 read 125 gammas. It was a relatively highly magnetized region.

Young found what he believed was a crystalline rock, the first he had seen. It was 12 centimeters long and one end resembled the triangular head of a diamond-back rattlesnake. The rock was white. When he held it up to the Sun, it exhibited a greenish-blue cast. Duke collected and documented two more samples of the Descartes formation material.

England then gave them updated magnetometer results. At the ALSEP site, the reading was actually 230 gammas, and at Spook Crater it was also 230, both record readings on the Moon.

Duke, who had fallen twice and was covered with dust, put a good-sized rock under the seat. He examined Young's crystalline rock and confirmed it was crystalline. According to premission guesstimates, the area should have been strewn with crystallized volcanics or plutonics. But nearly everything was breccia and a crystalline rock was a rarity.

They headed north in the Rover to Station 6, on the lowest terrace or "bench" of Stone Mountain. Looking northward, Duke told England he could see the lineations on Smoky Mountain, but they were more widely spaced than on Stone Mountain and mostly parallel to the Cayley Plain.

They turned westward along the terrace, trying to find a blocky crater, passing big, glass-spattered rocks that must have been thrown out of South Ray Crater. At length, they stopped at a 10-meter crater— a secondary one, they surmised, surrounded by big, angular blocks. They were scheduled to spend 20 minutes there, but were already 15 minutes behind their time line.

The Giants' Sandpile

Duke found a rock with a white matrix containing small, black crystals. Another breccia, Young said. Where were the igneous rocks they were commissioned to find?

Again, the orange juice dispenser entered the discussion. Young complained that every time he turned his head to read his oxygen gauge he got an ear full of orange juice.

At Station 6, they discovered the gnomon was broken and agreed they would use either the tongs or the shovel for scale in documented sample photos. Young dug into the ground with the shovel. He reported only soft, gray soil, not layered.

From Station 6 at the foot of Stone Mountain, the explorers turned the Rover westward and drove to Station 8 between two craters called Stubby and Wreck. There the Moonscape looked wild and tossed, like a sandpile where giants had played. Climbing a fairly steep ridge, the Rover began to lose power and Young found that the rear wheel drive had stopped working. They halted and he advised Houston that they may have severed a wire in the rear of the vehicle as they bounced over the rough surface. In any event, they had reached the area where it was expected they would find ejecta from South Ray Crater. They went to work. Duke began probing for a soft spot in which to take a double core sample.

"I don't know what this is staring up at me, Houston," said Young, "but I'm going to pick it up, because...."

"Anything that stares at you, you better pick it up," advised Tony England.

It was a big piece of glass that shone red, green, and rainbow colors in the sunlight. Young had found the first prism on the Moon. Duke reported there was a lot of glass, all over the area.

England asked the explorers to get as much variety in the samples as possible. Glass and breccia seemed to be all the region offered. When it was time to move on, England advised Young to test switch positions on the Rover in an effort to restore power to the rear wheels—each of which was turned independently by its own electric motor. A simple change of switch position brought the rear wheels back to life, but, as they drove northward down the ridge toward Stop 9, Young reported they had lost a rear fender. At this stop, they took a sample of undisturbed top grains of the regolith, using two types of nylon fabric like a dust rag to pick up the grainy material and store it in an aluminum box. This had to be done cautiously and they joked about "sneaking up" on the sample so as not to disturb the topmost layer. The "great lunar rock hunt," they called it.

They chipped a 4-centimeter piece off a boulder. The chip contained large, bluish crystals, some 5 millimeters long. Then they drove on to Station 10, near the lunar module, reporting ancient, subdued craters, some rimless, and a large depression with a hole in the center that looked like a slump.

Duke noticed that the navigation system had stopped working. It was no longer registering range or distance. At that point, the lunar module was not visible, but they headed for some blocks they could see atop Smoky Mountain to the north on the assumption the LM was in line with them. When they climbed a ridge, they saw the LM, dead ahead at 200 meters. Again, they marveled at how rough the area was. The only flat place seemed to be the one where they had landed.

Young made a soil test at Station 10 and asked for an extension of the EVA, which had already lasted nearly 7 hours. England said they were doing fine as it was.

"You said all we're gonna do tonight is sit around and talk," Young reminded him.

"Well, we like to hear you talk," said England.

"Yeah," said Young. "Especially on a hot mike."

They took another double core sample and Houston gave them a 10-minute extension, "just to show we love you."

"Atta boy!" cheered Duke. "Let's hear it for Old Flight! Atta boy!"

"Yea!" said Young.

One of their main tasks when they reached the LM was to shake off the dust covering them. They kicked a LM strut to get dust off their boots and banged each other on the back.

"Don't forget to kick your feeties," said Duke as Young started up the ladder.

Houston announced the EVA had lasted 7 hours, 23 minutes, and 26 seconds after they repressurized the LM.

"That's super," said Duke. "Let's go back out."

After two days exploring the Descartes region, the astronauts had a joint conclusion: "If this place had air," said Duke, "it sure would be beautiful."

The Third EVA

The objective of their third traverse Sunday, April 23, was North Ray Crater, about 4.9 kilometers from the landing site. They were allowed 40 minutes to get there and 1 hour and 5 minutes at the site. Because of their late landing, this final EVA was restricted to 5 hours—which included close-out time at the site. There were plenty of consumables left. But the explorers were rigidly bound by the time of lift-off, which had to be precise in order to make rendezvous and dock with Casper within the limits of the LM ascent engine and fuel supply. Mattingly was getting ready to change Casper's orbital plane on its forty-ninth revolution to compensate for the out-of-plane drift of the landing site with the Moon's slow rotation during the 3 days Orion was on the surface.

At North Ray, Houston scientists had anticipated the explorers would find volcanic rocks beneath the regolith and breccia. This expectation was shared by Young and Duke as they bounced over the boulder-strewn ground toward North Ray and saw huge boulders standing on its rim. The boulders grew in size as the employers approached the crater, climbing a steep slope toward its rim, which they had some difficulty in identifying. At one point, they reported they had reached the rim when they had topped a hummocky rise. It was just like mountain climbing, Tony England said—there is always another ridge.

When they reached the place they believed was the southeast rim, they found a level spot and parked. Young looked down into the crater. The inner slope was gentle for 50 meters, and then became steep.

"Man, does this thing have steep walls," he said. "They said 60 degrees. Now, I tell you I can't see to the bottom of it and I'm just as close to the edge as I'm going to get."

"That's the truth," Duke agreed.

When they aligned the high-gain antenna and turned on the television camera, England had to concur: "Man, is that a hole in the ground!"

Duke reported he could not see bedrock. All he saw were the boulders around the rim. The rocks on the ground were breccia, some with clasts of dark, glassy crystals. He and Young saw small rocks and pebbles as white as chalk. They collected them along with friable rocks with black clasts in them. Some of the interior black clasts had a distinct bluish tint. There had been green rocks at Hadley-Apennines. Now there were blue rocks at Descartes. Elsewhere on the highlands, there would be rocks that would appear to be orange—but that's getting ahead of the story.

On the rim, the regolith was soft and powdery. They sank down 6 inches as they raked up samples. Some of the breccia had a greenish tint to it, Young reported. There were black rocks here and there between the larger rocks. One enormous block seemed to be 20 meters long and 10 meters high—a huge breccia they called House Rock.

Young theorized that the black rocks came from the bottom of the crater and Duke picked up one he thought was "a typical basalt."

"Did you see anything you thought was igneous?" asked Tony England.

"This right here is an igneous rock," said Duke, handling the rock he called basalt.

"The whole place looks igneous, Houston," said Young.

But the overwhelming majority of sampled rocks were breccias. Duke noted these were of two types: one with black clasts in a white matrix and the other with white clasts in a black matrix. All the breccias showed signs of having been metamorphosized by shock. Several contained crystals of gray-white feldspar.

When they completed their collecting on the rim, they looked for a place that was permanently in shadow in order to take a special soil sample. It was surmised that small areas permanently in shadow might contain ancient volatile materials that would have boiled away in the Sun long ago. Chemists were curious about the nature of the volatile and semivolatile materials that might be trapped in cold, permanently shadowed soil. Young and Duke found a place they believed was permanently in shadow between House Rock and a 5-meter boulder at the south end of House Rock. Young reached in and scooped up the shadowed soil while Duke chipped off a piece of the boulder. He reported it had large vesicles in it—like drill holes or gas vent holes.

Leaving this site, they drove downhill, the slope seeming to be steeper going down than it had coming up. Houston called for a rake sample and a magnetometer reading. When the reading was taken, England announced it showed a local magnetic field of 313 gammas, a new lunar record.

"Well, the magnetic field of the Moon is a lot more than anybody ever believed it would be," said Young.

"That's right," said England. "From orbit, it looks like only 2 or 3 gammas."

They headed back to the landing site, detouring to the rim of a crater called Palmetto, to the west of their outbound route. From the rim, they estimated that the bottom was 100 meters, but they couldn't see it because of the steepness of the walls.

England asked Duke to collect samples of a vesicular basalt rock, which Duke had described earlier, but Duke was unable to find any. He said he might have led Houston astray by calling some rocks basalts when he wasn't sure of their type.

Back at the landing site, Young and Duke gathered up the aluminum-foil solar wind collector that they had mounted on its staff 3 days earlier, the cosmic ray experiment that had been monitoring solar and galactic nuclear particles, and the film pack exposed in the ultraviolet camera—three astronomy experiments for which the Moon was an ideal observatory.

By the time they had loaded their samples and equipment into Orion, dusted themselves off, and repressurized the cabin in preparation for lift-off, the third EVA had been clocked officially at 5 hours, 40 minutes, and 17 seconds. Their total EVA time was 20 hours, 14 minutes, and 55 seconds, nearly 7 hours less than planned, but still a new record. They had brought aboard 96 kilograms (211.2 pounds) of rocks and soil, compared with 77 kilograms on Apollo 15 and 21 kilograms on

Apollo 11. They had traversed 27 kilometers of the Descartes region and had been on the lunar surface 71 hours.

There had been numerous malfunctions of vehicles and equipment, but none seriously interfered with the mission. One of the final problems occurred as Young and Duke were doffing their Moon suits. Their helmets stuck and were hard to get off because of dried orange juice that had leaked out of the neck-band container.

At 7:15 P.M., Sunday, April 23, Houston signaled a "Go" for liftoff. The Rover television camera showed the ascent stage popping up in a cloud of dust like a champagne cork and over the audio system came a confirmation from the crew: "What a ride! What a ride! What a ride!"

Analysis showed that none of the samples Young and Duke returned supported the notion of a volcanic origin for this region.[3] As England later reported, the preflight, photogeologic interpretation of the Descartes-Cayley formation was wrong.[4] The big surprise was that breccias, not volcanics, were dominant.

Besides the white, crystalline pieces of impact-modified anorthosite, there was a species of breccia that was formed by the shearing and crushing of an ancient igneous complex in the region. It had white to dark gray fragments that were held together in a light gray matrix and was rich in anorthosite. This type of breccia had not been sampled on earlier Apollo missions, which returned breccias consisting of compacted regolith. The widespread anorthositic content tended to confirm the hypothesis of an early, anorthositic crust, especially since it was chemically related to the impact-modified anorthosite pieces.[5] There was a second type of breccia, light gray with many small fragments, which appeared to be compacted from the regolith, like the breccia returned on previous missions.

In addition to the modified anorthosite and the two breccia types there was a fourth rock type at Descartes—a tough, cohesive rock formed in a high temperature from molten material. The type was represented by the igneous, crystalline fragments Young and Duke found near North Ray Crater. Among the fragments were KREEP basalts, like those found at Fra Mauro. Their presence in the highlands suggested that these basalts, high in potassium, rare earths, and phosphorus, existed before the mare basins were filled and may have been part of a layer below the lunar crust.[6]

The absence of volcanism at Descartes was not the only surprise in the highlands. Another awaited explorers at Taurus-Littrow on Apollo 17.

Notes

1. Petrographic and Chemical Description of Samples from the Lunar Highlands, Apollo 16 Preliminary Science Report, NASA, 1972.
2. Third Lunar Science Conference, News Conference, Houston, January 13, 1972.
3. Preliminary Examination of Lunar Samples, Lunar Sample Analysis Team, Apollo 16 Preliminary Science Report, NASA, 1972.
4. England, Anthony W., "Summary of Scientific Results," Apollo 16 Preliminary Science Report, NASA, 1972.
5. Apollo 16 Rock Analysis Report, Manned Spacecraft Center, July 12, 1972.
6. Ibid.

12

Taurus-Littrow

The return of the first extensive collection of highland rocks by Apollo 16 dimmed the notion of extensive highland volcanism and offered instead new evidence that the highlands were very much older indeed than the maria.

The potassium–argon dating of the breccias gave ages in the range of 4 billion years. It appeared that except for erratic pieces of mare basalt thrown into the "terrae" by some energetic impact in a mare region, the Descartes-Cayley rocks were as old as any yet found on the Moon.

Oliver Schaeffer's group at the State University of New York, Stony Brook, dated one of the light matrix breccias (67483) at 4.25 billion years. It was the oldest rock found on the Moon, according to Schaeffer.[1] Generally, the Stony Brook laboratory found that the light matrix breccias from the vicinity of North Ray Crater were more than 4 billion years old. From South Ray there was a breccia that was dated at 4.08 billion years, some glass at 3.9 billion, and a fragment of anorthosite at 4.03 billion years.

These samples showed that rocks crystallized no longer than 300 million years after the formation of the Moon, Schaeffer concluded.[2] They showed something else, too, he pointed out—that considering the age of the youngest mare basalts found in Oceanus Procellarum was 3.2 billion years, it was clear that melting and crystallization processes on the Moon had been going on for a billion years, not merely 500 to 800 million years as indicated by the spread in ages of maria basalts.

Schaeffer's data together with magnetic, seismic, and heat flow results pointed to a Moon that had been thoroughly cooked in the period from 4.6 to 3.2 billion years and that might well be still simmering on the back burner.

Gerald Wasserburg concluded that the evidence from Apollo 14 and 16 and Luna 20 samples pointed to widespread melting and recrystallization of rocks between 3.8 and 3.95 billion years ago.[3] This period of 150 million years intrigued him. In every region of the Moon from which samples had been collected, there was evidence of cataclysmic bombardment in this period, Wasserburg concluded.

As I have mentioned earlier, there were several different kinds of ages that could be determined by isotope dating systems. The youngest age was the time of latest crystallization. It was evident early in the lunar dating game that some rocks were composed of materials that had crystallized and recrystallized several times. Nowhere was this rejuvenation process more apparent than in the breccia from Descartes.

These rocks consistently gave crystallization ages between 3.8 and 4 or 4.2 billion years, but they contained within their structure older components that were made 4.4 to 4.6 billion years ago. The greater ages, or "model" ages, revealed a primordial melting and crystallization at the supposed time the Moon was formed. For a period of 400 to 500 million years, these rocks remained intact. Then, in the period from 3.8 to 4 billion years ago, the primeval highland rocks were smashed up, melted again, and recrystallized. This surface rejuvenation appeared to Wasserburg to be a Moonwide occurrence, representing some global catastrophic period.

For example, he cited rock 12013, found in Oceanus Procellarum by the Apollo 12 team of Conrad and Bean. The crystallization age of the rock was 4 billion years, but there was evidence in it that it was derived from an earlier rock that crystallized 4.5 billion years ago. More than 900 miles to the southeast, at the Apollo 16 site, rock 65015, collected by Young and Duke, showed a crystallization age of 3.93 billion years, but it contained much older crystals that had not been remelted. The ancient crystals appeared to be as old as the chrondritic meteorites that are cited as markers for the period when condensation was taking place in the solar nebula. The crystals went back to the time of the formation of the Moon, or, at least, of its primeval crust.

With a double history, these rocks had a strange story to tell. Wasserburg concluded that at least part of it was a period of Moonwide catastrophe involving heating episodes and a tremendous bombardment of large bodies.

"The Moon," said Wasserburg, "is teaching us extraordinary things —like what the early history of a planet is like, a terrestrial planet, where we had no sense of formulating this prior to the information we derived from nature." In a sense, he pointed out, investigators were getting answers for which they had not formulated questions. He explained: "The complexity of the Moon is such that we could not have gone there with a pre-set list of questions. . . ." Lunar scientists were learning what questions to ask, but the lesson was late in the day, for the day of Apollo was nearly done. The last voyage, Apollo 17, was scheduled for December 1972.

The Stamp Scandal

In the summer of 1972, as it was beseeching Congress for funds to develop the space shuttle, NASA was rocked by a minor scandal. The principal victims were the Apollo 15 astronauts, Scott, Irwin, and Worden.

The scandal was triggered by the discovery that 100 postal covers they had carried to Hadley-Apennines had wound up in the hands of a West German stamp dealer who sold 99 of them for $1500 apiece. Since Project Mercury, spacecraft crews had been permitted to carry mementos of their voyages, such as medallions, postal covers, flags, and scarves in their personal preference kits or "ditty bags." But this was the first time that toting casual souvenirs to the Moon had approached a commercial enterprise.

A NASA investigation showed that the crew was persuaded to take the postal covers to Hadley by a promotion man at the stamp dealer's instigation. In return, the promoter established a $7000 savings account for each astronaut in a West German bank.

Each of the 100 covers carried a 10-cent commemorative Apollo 11 stamp that had been canceled at the Kennedy Space Center Post Office June 26, 1971, the day of the Apollo 15 liftoff; two 8-cent stamps that the crew bought and had canceled on their recovery ship, the U.S.S. *Okinawa*; the crew's autographs written on the envelopes during their flight from Hawaii back to Houston; and a notarized statement saying: "This is to certify that this cover was on board the Falcon at the Hadley-Apennine Moon July 30–August 2, 1971."

The astronauts subsequently declined to accept the money and returned to NASA 298 additional covers they took to the Moon on their own. The unauthorized covers were smuggled aboard Apollo 15 in the pocket of Scott's space suit, according to NASA.

The space agency also revealed that Scott carried two timepieces, a wrist watch and a stop watch, on the flight for evaluation, at the request of a friend. These items were not authorized, NASA said, and the name of the manufacturer was withheld to avoid commercial exploitation. The agency said Scott believed the timepieces might come in handy for the possible emergency timing of a manually controlled propulsion maneuver.

Another item carried to the Hadley-Apennines was a small statuette entitled "The Fallen Astronaut," which was planted at the site with a plaque commemorating astronauts and cosmonauts who had perished in the service of space flight development. The sculptor, identified as a Belgian named Paul Van Hoeydonck, subsequently donated a copy of the statuette to the Smithsonian Institution for display. Some time later, Scott learned that additional replicas were being offered for sale. The crew protested to Van Hoeydonck that commercializing the statuette violated the agreement under which the memorial was carried to the Moon, according to NASA. Nevertheless, 950 replicas of "The Fallen Astronaut" were offered for sale by a New York gallery at $750 apiece.

In the atmosphere of wheeling and dealing that has characterized government agency–industrial contractor relationships in the space age, the unauthorized freight that the Apollo 15 crew hauled to the Moon was a boyish prank. In the rhetoric of space program critics, however, it was branded as exploitation for personal gain of the most costly technological development in history. In the press, the astronauts were treated like fallen angels. "Say it ain't so, Dave," was the headline in one report.[4]

It was evident from the NASA investigation, which seemed to be reasonably thorough, that the only people who profited from the dereliction of the astronauts were the outsiders who put them up to it. The crewmen themselves received official, written reprimands that forever would mar their efficiency records as officers in the Air Force. What was especially damaging to their careers was a formal finding of lack of judgment. The punishment was the least NASA could do to appease the critics and a good deal more than many people in the agency considered appropriate. The astronauts themselves maintained a disciplined silence.

New Moon

On May 13, 1972, the seismometers on the Moon recorded the impact of a large meteorite on lunar far side. The recordings plus other data indicated the existence of a molten zone at the center of the Moon some 700 kilometers in radius because shear waves failed to pass through it. Shear waves do not propagate through a fluid and the zone was interpreted as a molten core.

The seismic evidence was supported by data from the Apollo 15 heat flow experiment and the electric conductivity profiles. These had been drawn from magnetometers on the surface and on Explorer 35, which was observing the Moon from orbit.

As the launch for Apollo 17 drew nigh, several lines of investigation were converging to offer a picture of the Moon—a New Moon, compared to the one that had existed in the minds of scientists before Apollo.

The New Moon had imported implications in the origin and evolution of planets. As Paul Gast described it:

> Prior to getting lunar samples, our concept of the chemical composition of the planets . . . was very simple minded. The bulk of the theories start with a simple-minded notion that the planets have overall chemistry like chondritic meteorites. I think we have now come to the point where we all recognize that we have a planet before us, the Moon, which is in no way like chondritic meteorites. In particular, it has very much more calcium, aluminum, uranium, thorium, strontium, barium and rare earths than any chondritic meteorite ever had by a factor of 10.
>
> That changed the whole concept of the composition of the planets. Because—if the Moon has that kind of composition, then we can throw out the idea that we know anything about the composition of Mercury or Venus in terms of chondritic meteorites. We have to look at each individual planet . . . and infer its chemical composition. A chemically heterogenous Solar System such as we are beginning to see from the Moon is certainly a fundamentally different Solar System from one where you basically make everything out of chondrites. . . .[5]

In terms of the Moon's emerging new image, the selection of a landing site for the last lunar mission was critical. Although all evidence so far indicated that the lunar heat engine had shut down 3 billion years ago, suggestions of more recent volcanism appeared in the Apollo orbital photographs.

One piece of significant photogeologic evidence was the presence of a dark mantling material in some highland areas. Not only did it appear to be of recent volcanic origin but it was much less cratered than adjacent regions and this attested to its relative youth.

The dark material was conspicuous in the Taurus Mountains southeast of Mare Serenitatis. Flying over the region in Apollo 15, Worden had taken high-resolution photos of the area and had reported seeing possible volcanic cinder cones.

NASA settled on a landfall for Apollo 17 in a valley no more than 6 miles wide in the mountain system southeast of the Serenitatis Basin. The floor of the valley was covered with the dark material that appeared to be volcanic. The closest major crater was Littrow.

The Taurus-Littrow site at 20 degrees, 9 minutes, 50.5 seconds north latitude and 30 degrees, 44 minutes, 58.3 seconds east longitude lay 450 miles east of Hadley-Apennines. Large massifs to the north and south and a hilly area called the Sculptured Hills to the northeast bordered the little valley.

Harrison H. (Jack) Schmitt, the lunar module pilot assigned to the mission and the only geologist to go to the Moon in Apollo, took some time out from preflight training to explain what he hoped to find at Taurus-Littrow.

"We are landing on a plain that seems to overlie a slightly elevated flat surface pre-dating the mare," he told a news conference.[6] "We have north and south massifs, peaks and mountains, which are part of the range structure of the Serenitatis Basin. Dark material mantles much of the area including the valley floor. It seems to be young. On Earth, it would be classic volcanic material."

Schmitt made the place sound like a geologist's paradise—as it undoubtedly was. There he hoped, he said, to find rocks younger than 3 billion years, possibly younger than 1 billion years. Was it too much to hope for, in view of all the evidence to the contrary on earlier missions?

Well, he said, there was a similar pattern of dark mantling material atop ejecta from the Crater Copernicus. If Copernicus was a billion years old, the dark material lying on top of its rays had to be younger— deposited later.

"It takes a big jump of the imagination in a sense to say, okay, the Taurus-Littrow dark covering materials are of comparable age, but it's a logical step to take," Schmitt said.

On the eve of the flight, Noel Hinners, deputy director of Apollo lunar exploration, commented that if, indeed, rocks at Taurus-Littrow turned out to be only a billion years old, "this would drastically alter our conception of the evolution of the Moon."[7]

The commander of the mission was Navy Captain Eugene A. Cernan, 37, who had flown on the 3-day mission of Gemini 9 with Thomas P. Stafford and had been lunar module pilot on the prelanding mission of Apollo 10 with Stafford and John Young. Cernan had logged 566 hours in space. A graduate of Purdue University in electrical engineering, Cernan held a Master of Science degree in Aeronautical Engineering from the U.S. Naval Postgraduate School. He had been awarded an Honorary Doctor of Laws degree from Western State University College of Laws and an Honorary Doctorate in Engineering from Purdue. He was married to the former Barbara Atchley of Houston and they had a daughter, Teresa Dawn, 8. In addition to the usual medals bestowed on astronauts after missions, Cernan had acquired an honorary lifetime membership in the American Federation of Radio and Television Artists.

Cernan's partner at Taurus-Littrow was Jack Schmitt, 36, who had been selected in 1965 as a scientist-astronaut and trained to fly at Williams Air Force Base, Arizona. A native of New Mexico, Schmitt received a Bachelor of Science degree from the California Institute of Technology and a doctorate in geology from Harvard University. He had spent a year studying at the University of Oslo, Norway, on a Fulbright Fellowship and had been a teaching fellow at Harvard.

One of the few bachelors in the astronaut corps, Schmitt served with the U.S. Geological Survey's Astrogeology Branch at Flagstaff, Arizona, before coming to Houston. He participated in photo and telescopic mapping of the Moon. He was the only astronaut who went to the Moon knowing exactly what he was looking for.

The command module pilot was Navy Commander Ronald E. Evans, 38, who had entered the Navy from a Navy ROTC program at the University of Kansas, where he was graduated with a Bachelor of Science degree in Electrical Engineering. After flight training, he had served as a combat flight instructor on two Pacific carrier cruises. Evans had a Master of Science degree in Aeronautical Engineering from the U.S. Naval Postgraduate School. He married the former Jan Pollom of Topeka and they had a daughter, Jaime D., 12, and a son, Jon P., 10.

When he was notified of his selection as an astronaut in 1966, Evans was on sea duty, flying F8 aircraft from the carrier, U.S.S. *Ticonderoga*, which happened to be the prime recovery ship for Apollo 17. The Navy is, after all, a small world.

The Weather from Orbit

Apollo 17 consisted of the command module America and the lunar module Challenger. The LM's famous namesake the H.M.S. *Challenger*, the British research vessel that laid the foundation of modern ocean science on a world cruise a century ago. Just 15 days short of 100 years after H.M.S. *Challenger* sailed out of Portsmouth, England, December 21, 1872, the lunar module Challenger was ready for launch from the coast of Florida on the night of December 6, 1972.

Liftoff of Apollo 17 was scheduled at 9:53 P.M., EST, the first night launch of the Apollo program. Residents from Miami to Jacksonville on the Florida coast and across the peninsula to Tampa–St. Petersburg were alerted to watch for the great pillar of flame from the Saturn 5 engines as the 363-foot rocket rose in the night. It would be visible for 250 miles.

Two minutes and 47 seconds before liftoff, the countdown was stopped. The sequence computer had failed to send the command to pressurize the liquid oxygen tank in the third stage (S4B) of the mighty Saturn rocket. The tank was pressurized with nitrogen just before launch so that the oxygen would flow into the engine to combine with hydrogen fuel when the time came to ignite the third stage hydrogen–oxygen engine.

In the Firing Room of the Kennedy Space Center, monitors corrected the omission by manually signaling the pressurization, but that wasn't good enough for the computer. Its logic circuitry insisted that launch preparation was not complete for the third stage. This would prevent the retracting of a gantry swing arm holding the rocket at T minus 30 seconds and the launch sequence would automatically cut off.

It was possible to bypass the faulty command circuit with a jumper. The trouble was that no one was sure whether the final 30 seconds of the launch sequence would work properly if the jumper was used. No one wanted to chance igniting the first stage engines and then having them cut off by the stubborn computer. That could delay the flight a month.

While the crew, Cernan, Evans, and Schmitt, waited in that grim, anxious silence that astronauts have endured during last-second count-down holds since Project Mercury, wondering whether it would be "Go" or "No Go" after all the preparation, the whole terminal sequence was painstakingly reviewed at the Marshall Space Flight Center, Huntsville. Marshall okayed installing the bypass jumper and the count was re-sumed.

Apollo 17 was launched at 12:33 A.M., 2 hours and 40 minutes late. It was December 7 by this time, another anniversary—the thirty-first since Pearl Harbor. The rocket rose on a pillar of orange fire that illuminated the pad and several square miles around it as brilliantly as daylight. As the rocket cleared the tower in the yellow-orange glare, a thunderous roar came back on the night wind from the Atlantic Ocean, shaking the buildings and the concrete-steel grandstands filled with news media observers, industry tycoons, and politicians and movie stars. The brilliant rocket exhaust made a high arc over the ocean, lighting up the water below. When the first stage engines shut down, the spectacle ended, and then a bright, distant spark signaled the ignition of the second stage. The last lunar Apollo mission was on its way.

During the Earth orbit phase of the journey and later during the long coast to the Moon, the astronauts described weather patterns they could see on Earth—in anticipation of the next manned space program, Skylab, on which weather would be closely observed from Earth orbit.

Jack Schmitt was particularly interested in watching the ice pack in the Ross Sea, an embayment in the gleaming ice cap of Antarctica. The crew reported a "tremendous clockwise rotational air mass" covering hundreds of square miles off Antarctica, near the tip of South America on the afternoon of December 9, as they were approaching the Moon.

Mission Control found their descriptions of Earth weather more detailed than weather satellite photos showing the same areas.

The Challenger landed in the valley of Taurus-Littrow at 1:55 P.M. December 11, and Cernan made the formal announcement: "Okay, Houston. The Challenger has landed." They had come down within 200 meters of the targeted landing point.

Shortly before 6 P.M., Houston time, Cernan and Schmitt stepped out on the surface to begin three hards days of exploration with an initial 7 hour, 12 minute EVA. They extracted the Rover from the base of the LM descent stage and tested it with a short drive around the landing site.

"Okay, Houston," reported Schmitt. "The basic material around the LM is . . . a fine grained, medium gray regolith . . . the craters bigger than about a meter in diameter seem to get to rock fragments." The geologist described the rocks he was seeing as pyroxene gabbros. They seemed to be rich in plagioclase, which characterized the anorthositic highland rocks from Fra Mauro to Descartes.

Cernan unpacked the remotely controlled television camera while Schmitt loaded the geology tools on the Rover. The dust was thick everywhere, and both men complained about it. But both were obviously enjoying themselves in their first hour outside on the Moon.

In fact, Cernan was so enthusiastic that the capsule communicator at Houston, Robert Parker, a scientist turned astronaut, warned that Cernan's exuberance was showing on the BTU (British Thermal Unit) measurements of his metabolic rate. The higher the BTU rate, the more oxygen was used, and oxygen was limited.

"Exuberance!" exclaimed Cernan. "I've never been calmer in my life!"

Then he added: "We'll take it easy, Rob. I think it's just getting accustomed to handling yourself in zero *G*. The only vice on the Moon."

"Rog," acknowledged Parker, an astronomer. "I thought you were at one-sixth *G*."

"Yes. You know where we are . . . whatever," said Cernan.

They completed setting up the high-gain antenna and as Cernan prepared to mount the color TV camera on the Rover, he called to Jack

Schmitt: "Oh, man. Just stop. You owe yourself 30 seconds to look up over the South Massif and look at Earth."

Busy with final antenna adjustment, Schmitt replied: "You've seen one Earth—you've seen 'em all."

"No you haven't," Cernan insisted. "When you begin to believe that. . . . C'mon, camera. Go in there. Okay! Camera is locked down."

"Hey, we have a picture, 17," said Parker. "We have a picture."

"Beautiful," said Cernan. "I hope it moves, now."

"It does," said Parker. He could identify Cernan in the bulky Moon suit because the mission commander was wearing a red band on his left arm.

During this exchange, Schmitt could be heard singing: "Oh, bury me not on the lone prairie, where the coyotes howl and the wind blows free."

They drove the staff for the flag into the ground with a few whacks from a hammer. The pole and flag shimmied like a tuning fork.

"Did you ever see a vibrator like that?" asked Schmitt.

"No. I've never put a flag up on the Moon before," said Cernan.

Cernan announced that the flag they had taken to the valley was the one that had been displayed in Mission Control throughout the program. Now, it was time to erect it on the Moon "in honor of all the people who have worked so hard to put us here," he said.

"Roger, 17," Parker responded. "We thank you very much."

"God, it's pretty up here," Cernan said.

The First EVA

New experiments had been added to the ALSEP on this flight and the explorers found most of their first outdoor period devoted to setting them up. This time the heat flow experiment was installed without accident, and later it returned results that confirmed the data from the instruments at Hadley-Apennines that a significant flow of heat was coming out of the interior of the Moon.

Two gravity meters (or gravimeters) were part of the package. One was implanted at the landing site to sense gravitational waves, if any, propagating through the universe. Similar experiments had been set up on Earth, with controversial results. Earth with its seismic and man-made noises made such a project difficult, but the quiet Moon was an ideal observatory.

In theory, gravitational waves emanating from the center of our galaxy would set up a characteristic vibration in the whole Moon. If such vibrations could be detected on the Moon at the same time as they were on the Earth, the existence of such waves would be demonstrated. Such evidence would go a long way toward the goal of unifying all the forces of nature: gravity, electromagnetism, and nuclear binding forces.

The heart of the gravimeter was a delicate spring balance that reacted to motions of the crust like a single-axis seismometer. It was sensitive enough to measure the bending of the crust by tides raised by the Earth and Sun. A portable gravimeter was carried on the traverse to measure subsurface structure by changes in gravity along the traverse route. Such changes would point to differences in rock masses below the lunar surface. An increase in mass at any locale would be detected by a higher gravimeter reading. This type of instrument had been widely used in prospecting for oil on Earth and in exploring the great ice desert of Antarctica.

The package also contained an atmosphere composition detector (a mass spectrometer) that recorded the density and composition in gas molecules at the lunar surface. Similar instruments were planted at the Apollo 15 and 16 sites.

Another new experiment was a boxlike instrument with two strips of thin film designed to measure tiny dust particles falling from space or ejected by meteorite impacts. When a particle struck the outer film, it produced an electrical signal that was repeated when it hit the inner film. The amplitude of and time between pulses indicated the mass and speed of a particle. The instrument would also indicate the particle's trajectory.

The Apollo 17 ALSEP carried a new, active seismic experiment. It was designed to draw a subsurface profile of the valley region. It consisted of an array of geophones that the astronauts laid out in the experimental area near the LM and 8 explosives that they planted along their traverse routes. The explosives were detonated by radio signals from Houston 24 to 75 hours after the crew left the surface.

In addition, the mission carried a radio transmitter, receiver, and antennas to probe the subsurface to a depth of a kilometer by means of high-frequency radio waves. The main object of this experiment was to determine if water or ice existed below the surface. The presence of water or ice on the Moon would virtually guarantee the development of scientific stations for long-term occupation. H_2O would provide oxygen for breathing, hydrogen for fuel and power, and water to sustain life, enabling a lunar colony to become chemically self-sufficient. Variations of radio waves at six different frequencies as they passed under and over the ground were recorded as a 300 to 3000 Hertz tone on casette tapes. The tapes were then returned to Earth for analysis.

Another new experiment was a neutron probe, emplaced in a drill hole during the first EVA and removed at the end of the third. The probe measured variations with depth in the soil in the rate of capture of neutrons produced by cosmic rays. The data were expected to provide clues to the depth at which lunar soil was mixed, the mixing history of cores the crew was collecting, and the depth at which rocks were irradiated by cosmic radiation.

A cosmic-ray detector was again carried in ALSEP. It was set to measure the mass and energy distribution of the solar wind and of cosmic-ray particles from the sun and other stars in the range of 1000 to 25 million electron volts.

In sum, the Apollo 17 ALSEP was the most sophisticated of all the scientific packages hauled to the Moon. It went much further than any previous array of experiments to demonstrate the usefulness of a lunar site as an outdoor laboratory for the observation of physical processes in the Solar System. The ground array was complemented by complex orbital experiments that Evans carried out in America. In every sense, the last Apollo mission not only attempted to explore a particular part of the Moon but sought to take the first step toward a lunar observatory. Apollo 17 was the end of the beginning.

Cernan and Schmitt spent so much time erecting these instruments, adjusting them, and making local observations that they fell 40 minutes behind schedule and were instructed to curtail their first traverse. Their abbreviated traverse was made due south of the LM for about 1 kilometer to the vicinity of a crater called Steno. They did not reach the crater's rim but were able to get samples of ejected, coarse blocks that seemed to represent the subfloor under the dark mantling material that covered the valley.

The surface around the landing site was a generally undulating, miniature plain. It had more blocks in the topsoil, however, and was considerably rougher than premission photos had led everyone to expect. The valley floor was saturated with small craters, most of them just a few centimeters in size, but boulders ranging from ½ to 4 meters were common. All of them were partly buried in or surrounded by dust from the dark mantle. The dark mantling material was thin. A small crater only 1 meter deep seemed to penetrate to the blocky subflooring below.

The dust was fierce. The site was probably the dustiest of any Apollo astronauts had visited. During the run in the Rover southward, both Cernan and Schmitt commented on the huge "rooster plume" of dust generated by the rear wheels. The plume not merely stretched behind the vehicle, but curved around to rain down on the men. At one stop, where they took a deep drill core, they discovered that the Rover had lost a rear fender on the right side. In spite of the dust, they collected 17 samples, including 3 rocks too big to fit in a bag with other samples, and 229 color and 197 black-and-white photos.

It was shortly after midnight, Houston time, December 11 when the astronauts returned to the Challenger. They spent a considerable amount of time dusting off each other before climbing up into the cabin.

Both Cernan and Schmitt advised Houston that they would have to do something about the missing fender. The dust kicked up was not acceptable, Cernan said. Joe Allen, the scientist-astronaut who was to relieve Parker as capsule communicator on the night shift, came on duty and promised to see what could be done. Obviously, the lunar explorers would have to improvise a fender if they hoped to make the kind of progress the time-line called for in the next 2 days. Parker told them they had been driving too slowly to complete the work laid out for the second EVA. Cernan explained that they had done a good deal of circling. Besides, the dust was a major handicap.

Parker stayed overtime to debrief the crew on the first EVA. He said the Houston scientists were still puzzled about the dark mantle.

"There has been a lot of discussion today about whether or not it could have been a regolith derived from the intermediate gabbro which you were sampling from a boulder," Parker reported.

Schmitt replied, "I do not have an intuitive feeling that the regolith has been derived from most of the boulders we're seeing, because those boulders are fairly light colored. They look like they're probably 50 percent plagioclase."

As it turned out, Schmitt's "intuition" was right. The eventual determination of the origin of that dark mantling material was to be a surprise. Cernan said he felt that the boulders they were seeing represented the subfloor of the valley.

In the mountains, Schmitt said, there appeared to be linear structures. The lineaments dipped toward the southwest in the great pile called the South Massif. Not far away, another pile, called Bear Mountain, seemed to have the same kind of "organization," but Schmitt said he could not be sure because the Sun-angle was wrong for clear observation there.

Summing up, said Parker, it was the consensus at Houston that an intermediate, medium-grained, vesicular gabbro formed the upper layer of the subfloor. The dark mantle composition remained speculative at the time.

A Paper Fender

Early in the morning of December 12 at Houston, John Young put on a space suit and tried to rig a fender made out of map paper on a mock-up Rover at the Johnson Space Center. He wore the space suit to see if the fix could be made by space suited people on the Moon.

When he completed the experiment, he doffed the suit, marched into Mission Control and called up Cernan and Schmitt, who were finishing their rehydrated breakfast in the Challenger.

The procedure to rig a replacement fender, he said, is to take four pages out of the lunar surface map book, tape them together with gray tape, and construct a large piece of heavy paper 15 inches long by 10½ inches wide. This could be laid on top of the Rover's fender guide rails and clamped to the rails with clamps from a portable utility lamp. Young estimated it would take only 2 minutes to perform the clamping. This was better than spending 12 minutes or more dusting each other off after a traverse, he reckoned.

Cernan agreed the scheme might work. Young then told him to allow about an inch of overlap and to make certain he squeezed all the oxygen bubbles out of the taped pieces—otherwise, as soon as the assembled map-fender was taken outside the 5 pounds-per-square-inch atmosphere pressure of the Challenger's cabin, gas bubbles would expand into the lunar vacuum and blow the construction apart. It was a tricky job, but Cernan said it was better than nothing.

Cernan and Schmitt went to work and soon had taped together four sheets from the map folio, making a large sheet actually 15-by-19 inches. At 5:30 P.M., CST, December 12, Cernan and Schmitt climbed down from the Challenger with their map-fender rolled up in an equipment bag and proceeded to the Rover to make the repair. About a half dozen fellow astronauts were clustered around the capsule communicator's console at Houston to observe this operation on television.

"Nice day," said Schmitt, conversationally. "Not a cloud in the sky."

The Johnson Space Center Public Affairs Officer offered an interpretive comment: "As you can see, it's only a paper fender, but the Moon is real."

The taped paper held together and the clamps fastened it on the fender guide beautifully. Standing back to survey their work, the explorers were conditionally satisfied with the effort. They had undoubtedly plowed new ground in space flight; history might remember that a multimillion dollar vehicle had been fixed with paper, tape, and clamps. Paper clips had been considered, but were rejected as too weak. Now the explorers were ready to go on their second traverse.

The first stop was designated as station 2, about 7 kilometers west southwest of the Challenger, on the flank of the huge sandpile called the South Massif.

The Sun was higher than the previous day and the glare stronger. Mission Control kept a close check on their progress by means of Schmitt's descriptions of landmarks. Parker called up to them with the reminder they were limited to 63 minutes to get to Station 2.

Schmitt reported that the surface was not changing much in detail. The fine-grained texture of the regolith continued westward. Micro-meteoroid impacts made a raindrop pattern in it and the big blocks looked the same as those sampled the day before. There were a number of small, whitish craters, each of which had a pit in the center lined with glass.

"Boy, am I glad we got that fender on," said Cernan. The paper fender was working like a charm, diverting the "rooster tail" of dust behind the vehicle.

Cernan observed that the Rover must be climbing because the best speed he could get out of it was 10 kilometers per hour "at full bore." Bronte crater loomed up bigger than expected and Cernan had to detour its rim. Beyond, Schmitt observed several bright craters with dark halos around their rims. Such halos signified a past volcanic eruption in similar craters on Earth. However, a volcanic interpretation of the fresh-looking craters in the valley was weakened by the close-up inspection of the dark mantle.

Schmitt was coming to the conclusion that the mantling material was not recently volcanic, but it was dark in color because of the presence of ilmenite, an iron-rich mineral. Another possibility, he mentioned, was that the dark soil contained fine glass that was blackened by iron and titanium, like the Apollo 11 basalts from Mare Tranquillitatis. It was apparent to the geologist that the dark material was not the ash and cinders spewed out from a volcano that the premission photos had suggested. No ash or cinders were ever returned from the Moon.

As they approached the South Massif, Schmitt reported that the lower slopes were strewn with blocks. Looking across the valley at the North Massif, Cernan said he could see "wrinkles" on its flank. Schmitt confirmed this, saying there was no question that there were lineations all over these massifs in a variety of directions.

During this trip, Ron Evans passed overhead in America, performing infrared, ultraviolet, and gamma-ray observations of the surface from the well-equipped command and service module. On Apollo 17, the spacecraft, America, was outfitted with sensing instruments like a lunar version of Skylab. There was no doubt that the scientific community

APOLLO 🅸🅷 TRAVERSES

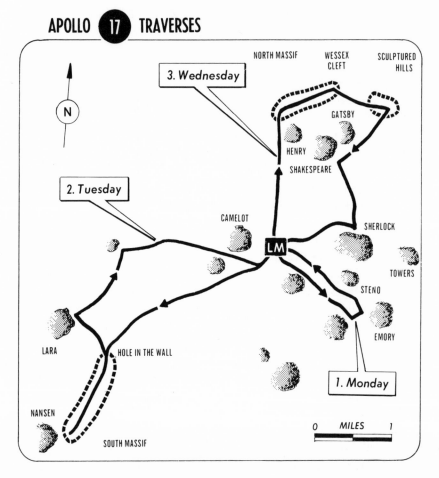

NORTH MASSIF WESSEX CLEFT SCULPTURED HILLS

3. Wednesday

N

GATSBY

HENRY
SHAKESPEARE

2. Tuesday

CAMELOT

SHERLOCK

LM

TOWERS

STENO

EMORY

LARA HOLE IN THE WALL

1. Monday

NANSEN

SOUTH MASSIF

0   MILES   1

was squeezing every possible kind of datum out of the mission. It was the last chance at the Moon for a long time.

The explorers cruised along the flank of South Massif and then parked at Station 2, where they turned on the television. Parker asked them how the paper fender had held up. Beautifully, they said. They collected samples and Schmitt gave a gravimeter reading. Cernan picked up a crystalline rock he thought was a porphory, another igneous rock containing feldspar. It was bordered with reflecting crystals. From one 20-inch boulder, a blue-gray breccia that seemed to have rolled part way down the South Massif, Schmitt and Cernan collected a large clast of dunite, an ancient igneous rock usually associated with the mantle. Poking around, Cernan picked up a medium green anorthositic gabbro nearby. It looked as though it had pastel green crystals in it. He put a chip of it in the sample bag.

Urged on by Houston, they turned north and drove to the vicinity of a crater called Lara, which was Station 3. They paused to take a core sample and collect some blue gray breccia, which was strewn around the rim of the crater. Urged on by Mission Control, they took off toward the northeast, toward a crater called Shorty. It was Station 4. En route, Schmitt reported he was getting a good view of North Massif. The lineaments now appeared to be cross-hatched. One set plunged 30 degrees downward to the east and the other, 30 degrees downward to the west. There were boulder tracks on the mountain, where boulders had rolled down.

The Orange Soil

Shorty was a "fresh" impact crater, about 110 meters in diameter, with a dark rim and a jagged, central peak. The first thing the explorers saw as they parked on the rim was a big, intensely fractured boulder. Schmitt said it looked like clinopyroxene gabbro, an igneous rock, and was obviously crystalline.

Although the rocks they had seen were all igneous, products of lunar cooking, they did not represent any recent volcanic activity. They appeared to be very old, but they were undoubtedly products of an early melting episode that had created wide areas of the highlands.

SCHMITT: Oh, hey! There is orange soil!
CERNAN: Well, don't move it till I see it.
SCHMITT: I stirred it up with my feet.
CERNAN: Hey, here it is. I can see it from here.
SCHMITT: It's orange.
CERNAN: Wait a minute. Let me put my visor up. It's still orange.
SCHMITT: Sure is. Crazy! Orange!
CERNAN: I've got to dig a trench, Houston.
HOUSTON: Copy that. I guess we'd better work fast. [Not much time had been allotted to Station 4.]

The discovery of orange soil on the Moon signified the probability of recent volcanic activity. If so, it was what the crew had come to the Taurus-Littrow region to find. Word of the orange soil spread that evening through the motels around the Johnson Space Center. Scientists began calling the Center from their rooms to find out what was going on. Orange soil! It was like discovering a new planet.

Schmitt told Parker as Cernan began to dig the trench: "He's not going out of his wits. It really is orange. It's almost the same color as the LMP (lunar module pilot) decal on my camera."

CERNAN: That is orange, Jack.
SCHMITT: I didn't think there would be orange soil on the Moon.

It looked to Cernan like an oxidized desert soil. The crater, then, began to take on the appearance of a volcanic fumerole. If this stood up as evidence of recent volcanism, the scientists who insisted that the Moon's heat engine shut off 3 billion years ago were on the wrong track. The whole conception of the Moon's evolution would have to be revised.

Cernan feverishly dug the trench across a layer of the orange soil that seemed to be about 1 meter wide and extended around the rim of the crater. He scooped samples of orange soil out of the trench and samples of gray soil on either side of the orange band.

Houston called to warn the explorers that they were running into "walkback constraints." They had 20 minutes left before it would be too late for them to reach the LM safely on foot if the Rover broke down. Time and distance of the traverses were dictated by the ability of the explorers to return to the LM on foot within the limits of their oxygen supply.

Cernan reported that part of the orange soil zone was actually crimson in color and faded to orange and then gray.

They took a gravimeter reading and drove a core into the orange soil, deep. It took a good deal of tugging to get the core out, but when they did, they saw that the soil on the core tube was red in the center, black at the bottom and orange on top, with some admixture of gray soil.

"Fourteen minutes," Houston was counting down to departure. "Thirteen now. . . ."

Cernan noted that the orange soil continued down radially into the crater. There was a lot of it.

At that point, there was considerable excitement at the space center, where it was agreed that Cernan and Schmitt had made one of the major finds of the Apollo program.

As they moved on to Station 5, Cernan remarked that Shorty Crater could have been a volcanic vent, "but we didn't have time to prove it." En route to Station 6, he set up another charge of explosive for the seismic profile.

In a report on the discovery, the U.S. Geological Survey noted the next day that the red-orange soil layer seemed to run around the circumference of the crater and resembled "alteration halos which occur around many terrestrial volcanic vents." But the USGS report added cautiously that the "morphology of Shorty, however, is similar in some respects to impact craters."[8]

Although a "definitive interpretation of its origin may depend on sample analysis," the report stated, "this crater appears to be volcanic, thus suggesting volcanism before and after formation of the light mantle material." The light mantle, it added, appeared to be derived from the massifs, but the origin of the dark mantle was not yet established.

Cernan and Schmitt drove hard to reach the LM area on schedule. Their EVA had lasted 7 hours, 37 minutes, and 22 seconds, a new outdoor lunar record.

Meanwhile, Mission Control passed the news about the orange soil up to Ron Evans in America as he prepared to retire for the night: "While you're doing your presleep checklist, you may be interested that at Shorty, the surface crew found some very, very orange soil, a great deal of it. Indicates strong oxidation and possibly indicates water and/or volcanics in the area. And they're really—Jack is kind of like a boy at Christmas time. I'll tell you. A little kid at Christmas time on that one."

"I bet he would be!" said Evans, laughing. "Hey, that's a great find, by gosh."

"Yes, that's the first time we find . . . it's orange. Boy, you could see it in the television; it's bright orange soil. No question about it," said Houston. "And as luck would have it, they found it and got working and then they had to pull out of Shorty due to constraints—walkback constraints in the area. . . . You know, consumables versus walkback."

"Yes," said Evans.

The Final EVA

On the third and last EVA of the mission, and of Apollo on the Moon, Cernan and Schmitt motored northward to the North Massif on the evening of December 13. Cernan described fields of boulders. On the massif, he and Schmitt could see a line of boulders about 100 meters long. From that array there were downhill tracks, showing that some boulders in the mass that had fallen there had rolled on down the hill.

Parker advised them to keep a sharp eye out for more volcanic evidence, especially near a landmark called Van Serg Crater, southeast of the mountain. They passed along the uplifted domes called the Sculptured Hills, which Schmitt noted were different from the mountains in structure, with subfloor types of rocks. They paused at the Van Serg Crater long enough to see that it had boulders in the center, forming a mound about 30 meters in diameter.

"I don't see any orange material," said Cernan. Neither did Schmitt.

They scooped up samples—glassy fragments, rock fragments, and one rounded, baseball-sized rock covered with glass.

Cernan examined the paper fender and reported sadly that it was almost worn out. Well, they were almost through on the Moon, anyway. They headed back to the Challenger.

As Ron Evans passed by overhead, Houston asked him to take a good look at Shorty Crater.

When the ALSEP experiments had been checked and all the material to be returned to Earth had been collected, Cernan unveiled a plaque affixed to a leg of the Challenger's descent stage, which would remain on the Moon.

The plaque showed a map of the Earth and a pictorial view of lunar near side, showing all the six Apollo landing sites. It was left there, Cernan said, "so that when this plaque is seen again by others who come, they will know where it all started."

The Analysis

In total, Apollo 17 returned 335 rocks, 73 clumps of soil, 8 cores, a deep drill sample, and pieces of boulders to the Lunar Receiving Laboratory at Houston. The total weight was 110 kilograms. Analysis of the orange soil was eagerly awaited by the lunar science community.

The Preliminary Examination Team found that the soil consisted of ruby red to black spheres and broken spheres of glass, with a trace of olivine phenocrysts. The black portion contained ilmenite. It was not of recent volcanic origin, not by a long shot. It was dated at 3.7 to 3.8 billion years old.[9] The dark mantling material in the valley of Taurus-Littrow turned out to be pulverized mare basalt, similar to that collected by Armstrong and Aldrin on Apollo 11 at Tranquility Base. It was basalt with a high iron and titanium content, which gave it its dark color. It also dated as 3.7 to 3.8 billion years of age.

"Evidently, widespread volcanism involving melts rich in titanium occurred over much of the Moon's eastern limb 3.7 to 3.8 billion years ago," the report said. "Thus the known time span of mare volcanism remains essentially the same as before the Apollo 17 mission."

From Apollo 11 to Apollo 17, the investigation of the Moon's volcanic history had come full circle.

The New Conception

With the end of Apollo, a new conception of the origin and evolution of the Moon began to take shape in the predominant view of lunar scientists.

A New Moon rose at the Fifth Lunar Science Conference at the Johnson Space Center, Houston in March 1974. It was a planet in its own right, with a thick crust, a mantle, and an iron or iron sulfide core. It had been formed in the inner Solar System from a condensate that was low in volatile elements and high in those that are refractory—except from iron and siderophile elements. Such a planet had to be captured to become a satellite of the Earth.

Inasmuch as the development of a scientific theory is a process of induction that is highly susceptible to modification as new data accumulate, the New Moon can be regarded only as a temporary model. New Moons were on the horizon throughout the entire Apollo period. Some formed in Earth orbit from debris left over from the formation of the Earth or from a primeval silicate atmosphere around the Earth. Others were created by fissioning from the Earth at several stages of its formation or by accreting in a double planet system with the Earth.

This much can be said for the Moon of 1974. It grew out of the first conference that provided an overview of the six Apollo landings. The Fourth Conference in March 1973 followed Apollo 17 too closely to allow time for an assessment of its results.

The New Moon left many questions unresolved, for Apollo was far too limited in time, in scope, and in equipment to do much more than scratch the surface of the fundamental questions of the creation, which are the fundamental questions about the Moon. But what had been achieved in the voyages of Apollo was the establishment of physical and chemical constraints into which any theory of lunar origin and evolution must fit.

On this intellectual Procrustean bed, the New Moon was born, the Moon of Apollo. It was the most probable model from Apollo data.

An early indication of the Moon's individuality was its difference in bulk composition and in its ratio of critical chemical elements and their isotopes from the Earth and from meteorites. There was, for example, the Moon's lack of water, which denied the development of life, its relatively high abundance of uranium (59 parts per billion) compared with the Earth (18 parts per billion), and with chondritic meteorites (11 parts per billion), and its lower potassium-to-uranium ratio than that of the Earth. These clues all pointed to a separate origin.

At the Fifth Conference, Edward Anders and an associate, R. Ganapathy, of the University of Chicago proposed that planets and chondritic meteorites had a bulk composition of varying mixtures of six basic components: an early condensate, a remelted silicate, and unremelted silicate, a volatile-rich silicate, troilite (iron sulfide) and metals.

Condensation took place in a chemical sequence from the solar gas.

The early condensate was cooked in the high temperature of the nebula, separating the silicates from the metals. Later remelting produced further fractionation of the material.

"God used the same ingredients to make the Moon as the chondrites," Anders commented, "but the proportions were different."[10]

As Paul Gast had observed in 1972, the difference in the recipe for the Moon and for meteorites was a fundamental discovery in Apollo. And yet, as the Chicago scientists noted, there was a familial resemblance among the bodies of the Solar System. This was pointed up by three other University of Chicago scientists, Robert N. Clayton, Lawrence Grossman, and Toshiko Mayeda. They had found a grain of interstellar material in an analysis of the famous Allende Meteorite. The foreign material was characterized by an anomalous amount of pure oxygen-16, an isotope normally mixed with oxygen-17 and 18 in all Solar System material analyzed. The presence of the unmixed oxygen-16 suggested that the alien particle originated in a primitive star, rather than the Sun, which is a third- or fourth-generation star, and somehow survived in the solar nebula to become an erratic inclusion in the meteorite.[11]

In other respects, the composition of the Allende Meteorite was similar to that of lunar soil. Whitish nodules in the meteorite were enriched in refractory elements that would condense at the highest solar gas temperatures and in rare earth elements like the KREEP basalts.

The Taylor-Jakes Model

It was predictable that with the data from six missions, a theory of lunar structure would be proposed in keeping with the chemical, seismic, magnetic and heat flow results, and the observations from orbit. And it was—by S. R. Taylor of the Australian National University and P. Jakes of the Geological Survey of Czechoslovakia.[12]

They surmised that the Moon accreted in a part of the solar nebula where the condensate was enriched in refractories, except for iron and siderophiles, and depleted in volatiles. Heat produced by accretion caused the heavier elements to move to the center and the lighter ones to rise to the surface. At least 80 to 90 percent of the Moon must have been melted to allow such a concentric, layered structure of a core, mantle, and crust.

This model seemed to be consistent with the Apollo data, but before Apollo, it would have been summarily rejected. In pre-Apollo thinking, two observations precluded a metallic core. First, the Moon's overall density seemed to be too low. Second, the moment of inertia ratio

(I/MR^2) appeared to be that of a homogenous sphere. This ratio (0.4) was estimated by telescopic measurement of the Moon's sideways oscillation or liberation. It suggested strongly that the Moon was a homogenous chunk with the density of the Earth's mantle, a concept that was one of the mainstays of the fission theory. It precluded a concentration of mass at the center.

A planet like the Earth, with a heavy, iron core, would have a lower moment-of-inertia ratio; a hollow sphere with mass concentrated toward the surface would have a higher one.

After laser mirrors had been set up in Mare Tranquillitatis, Fra Mauro, and the Hadley-Apennines, it was possible to measure liberation more precisely with ruby laser beams from the McDonald Observatory. Using these data, W. M. Kaula of the University of California at Los Angeles and his associates calculated the moment of inertia ratio as 0.3953.[13] Because this result was lower than that of a homogeneous sphere, it allowed some high-density material to exist in the center. The refined ratio was consistent with an iron or iron sulfide core.

If the core was iron, investigators estimated, it would be 500 kilometers in radius. If it was iron sulfide, it would be larger—about 700 kilometers in radius. On the basis of these estimates, the Moon could have a metallic core of 28.7 to 40.2 percent of its mean radius, which Kaula and his group determined from laser altimeter measurements to be 1,738.2 kilometers.

The invocation of a near molten Moon at the beginning by Taylor and Jakes had been hotly disputed in the past, especially by Urey. By the time of the Fifth Conference, however, resistance to primordial melting had weakened. It seemed to be difficult to explain highland structure otherwise.

A missing link in the evidence of early melting had been a true genesis rock, one which could be shown to have crystallized from a melt 4.6 billion years ago. With a few older and younger exceptions, the ages of the highland breccia clustered around 4 billion. None had been found that had crystallized at the time of the Moon's formation up to Apollo 17.

It was considered probable in 1972 that none would be found because the chemical isotope clocks had been stopped or reset by impacts in the highlands 4 billion years ago.

Candidates for the title of genesis rock had been scooped up at Fra Mauro and at Hadley-Apennines. They had turned out to be less than the age of lunar creation. Was there a real genesis rock?

Yes there was. It finally turned up on Apollo 17, the last chance to find one. It was the dunite clast that Cernan and Schmitt collected from a blue-gray breccia boulder at Station 2 on the flank of the South Massif.

Gerald Wasserburg and his colleagues at the California Institute of Technology dated a chip of dunite rock as 4.6 billion years old by the rubidium–strontium system. Much of the clast had somehow survived high shock pressures and heat that deformed and recrystallized parts of it and eventually crushed it into the breccia.[14]

The piece contained large crystals of pale green, translucent olivine in a granular white matrix. It had been excavated at great depth by one or more impacts, probably from the mantle.

The genesis rock not merely fulfilled half of Schmitt's hope to find the oldest and youngest rocks on the Moon at Taurus-Littrow, but offered proof of primordial melting that could hardly be dismissed.

In addition, Taylor and Jakes cited the presence of uranium and theorium near the surface as evidence of deep, primordial melting. Convection, they surmised, had brought the uranium and thorium upward from the deep interior. This process was abetted by the progressive crystallization of such minerals as olivine, which pushed the radioactive elements surfaceward as convection cooled the interior.

Thus, Taylor and Jakes accounted for the concentration of radioactive elements in the upper Moon, 100 to 300 kilometers below the surface.

This arrangement set the stage for the second melting period, starting about 700 million years after the formation of the Moon and continuing for 800 million years: from 3.9 to 3.1 billion years ago. The relatively high abundance of radioactive elements in the upper Moon, as indicated by the heat flow and gamma-ray data, provided the radiogenic heat for this episode, which created the maria basalts.

Hell Broke Loose

In sketching the development of the 1974 model to this point, I must backtrack a hundred million years or so to report what investigators surmised about the basins. Wasserburg, as mentioned earlier, proposed a period of catastrophic bombardment between 3.85 or 3.9 and 4 billion years ago as the mechanism for basin formation. During a span of 100 million years, he said, all hell broke loose on the Moon. It was plastered with rocky planetesimals. These have a role in two other theories I will describe in a moment.

Wasserburg asserted that his data proved conclusively that there was widespread redistribution of lead and uranium all over the near side between 3.9 and 4 billion years ago.

"This observation in conjunction with the sharp cut-off in crystallization ages at 4 AE [aeons or billions of years] . . . provides strong evidence of a terminal lunar cataclysm. . . . This cataclysmic event or cluster of events which may have occurred over a 0.1 AE interval (100 million years) is most reasonably associated with the formation of major lunar basins and temporally related magmatism."[15]

Wasserburg's terminal cataclysm impressed many of his colleagues at the Fifth Conference. Many seemed to accept it, but there was dissent from Oliver Schaeffer's group at the State University of New York, Stony Brook. For some time, there had been a low-keyed rivalry between the "asylum" in Pasadena and the laboratory on Long Island. The two laboratories seemed to be competing in the lunar rock dating game, and they were two of the best in the world. On the question of the period of the cataclysm, however, they were more than 3,000 miles apart. They were 100 million years apart.

Schaeffer and Liaquat Husain maintained that basin formation was spread over hundreds of millions of years and did not occur in a relatively brief period near the marker of 4 billion years ago—the time of the great Imbrium impact.[16] The cluster of rock ages at 4 billion years old could be explained by the probability that all the rocks of that age came out of Imbrium, whose ejecta was spread all over the near side.

The impact had reset the chemical isotope clocks, Schaeffer and Husain reasoned. Heat and pressure had caused degassing of argon from the rocks, for example, thus resetting the potassium–argon timing system. As the rocks (mostly breccias) cooled, the clocks had restarted. Argon began to accumulate from the decay of potassium-40. When man came, the clock said 4 billion years—the time of the Imbrium impact.

Using Imbrium as a starting point, the scientists traced other samples to other basins that then appeared to be younger or older than Imbrium: Mare Orientale was 3.85 billion years old; Mare Crisium, 4.13; Mare Humorum, 4.13 to 4.2; Mare Nectaris, 4.2.

"It would appear, then, that after the lunar crust formed, there was an era of multi-ring basin formation lasting until 3.9 GY [giga years, another term for a billion years]," they said. "This era was followed by an era of basalt flooding of basins and low lying areas on the lunar surface, which lasted to 3.1 GY. At this time, all major activity ceased."

Did it? Schmitt was not convinced of that when he went to the Taurus-Littrow Valley to seek evidence of younger volcanism, only to discover that he had been tricked by the camouflage of the dark mantling material. Nor was James W. Head of Brown University, who had made an analysis of the dark mantling material.

Head told a news conference that although the dark mantle did not represent recent volcanism at all, there were indications of lava flows younger than 3.1 billion years in the western region of Oceanus Procellarum—where no landings took place. It is possible, he said, that melting continued in the western region longer than in the central and eastern sections where Apollo missions did land. In the orbital photos lava flows in western Procellarum clearly overlapped those in the eastern part, he said. They must have come later. Also, he added, crater densities were lower in the west. Unhappily, for the foreseeable future, this possibility would have to remain speculation. There was no way to get there at least until the twenty-first century, unless Russia flew a grab sampler to the region.

Head's analysis of the dark mantling material showed that it was a mixture of basaltic fragments and spheres of orange and black glass.[17] The black glass gave the mantle its dark color, which had lured Apollo 17 to the region.

Head surmised that the deposit was formed in lava fountains erupting from the crust in the volcanic outburst that produced the basaltic subfloor of the valley 3.63 to 3.83 billion years ago. It must have been a spectacular period for, if Head was right, there would have been spectacular lava fountains rising all over the Moon. The dark mantling material was widespread. It was observed from orbit in regions southeast of Mare Vaporum, southeast of Sinus Aestuum, southeast of Copernicus, on the floor of the Crater Alphonsus, at the south-southwest edge of Mare Humorum, on the Aristarchus Plateau, and in north central Procellarum around Schröter's Valley.

Some of the sites—Alphonsus, Artistarchus, and Schröter's Valley particularly—were places where the mysterious mists and fogs had been reported by terrestrial Moonwatchers. The sites also had registered higher gamma radiation than surrounding areas, indicating higher radioactivity.

Orange and Green Glass

Taurus-Littrow had one more surprise—the significance of the orange soil. Its color was derived from glass spherules, some black and others ranging from deep red to light brown. These were also presumed to be products of lava fountaining, as noted earlier.

But of Moon-shaking importance was the discovery that clinging to the surfaces of the glass were volatile elements—in which other areas of the Moon were woefully deficient. There are halogens,* especially bromine, and silver, zinc, cadmium, and thallium.

* Halogens are a family of active chemical elements consisting of fluorine, chlorine, bromine, iodine and astatine.

Not only was the orange glass enriched in associated volatile elements, but it was depleted in refractory elements—the reverse of the element abundances found in earlier samples. It appeared as though the glass represented material that was complementary to the supposed bulk composition of the Moon.

Was the glass imported by meteorites? No, it was certainly indigenous, according to Anders.[18]

The possibility was considered that it was a clue to some undiscovered concentration of volatile elements deep within the Moon or on the far side. Only a fraction of lunar near side had been explored. Determining its nature from Apollo samples was analogous to making a chemical survey of Africa from six sites near the equatorial part. The existence of a large concentration of volatiles would probably cause the New Moon to set.

The conclusion that the orange glass was indigenous was supported by Ralph Baldwin, an industrialist as well as a scientist, who had been an authority on the Moon long before Apollo. The glass was much too widespread to be of meteoritic origin, he told a news conference. It was undoubtedly a product of lava fountaining, he said, with magma spouting up from surface vents under pressure—gas pressure. The source of the volatiles on the glass might be condensed gases, he suggested.

Some implications of the volatiles were spelled out by Leon Silver at the news conference. He was a conservatively imaginative chemist and his words carried weight. In his view the volatiles created a doubt about the origin of the Moon from a volatile-depleted condensate. Perhaps the Moon was not as depleted as supposed in volatile elements, perhaps the assumption that it was failed to take into account the existence of volatile zones, perhaps the Moon the scientists thought they were seeing was not the real Moon at all, perhaps they were being fooled, as the early sixteenth-century mariners were when they believed that Newfoundland was a promontory of the coast of China.

The issue of the volatiles had become a serious consideration. Silver explained:

> People have gone back and looked at the green rocks of Apollo 15. They are composed of small spherules of green glass, similar in their bulk chemistry and in volatiles to the orange glass.
>
> These volatile enrichments indicate the presence somewhere in the Moon of abundances of volatiles, comparable to solar abundances and similar to the Earth abundances.
>
> [If the glasses actually came from the Moon] it would then appear that the Moon may not have accreted in precisely the form that we first inferred. Some ardent advocates are backing away from a conclu-

sion of a simple, high-temperature condensate accretion. There may be more volatile materials on the Moon than we thought.

The question was highly significant, Silver maintained, for it might well influence the whole approach to further geochemical exploration of the Moon—if any further exploration there was to be.

The orange glass and the green glass and the black glass—all bid fair to become an embarrassment to the theory of accretion from a volatile poor condensate in the solar nebula. Hardly born, the New Moon already was in some trouble.

The Resurrected Moon

There was another problem in the theory—the low siderophile abundance. What happened to the iron, nickel, gold, and platinum?

A solution was provided by two new theories of lunar capture presented at the Fifth Conference, one by Joseph V. Smith of the University of Chicago[19] and the other by John A. Wood and H. E. Mitler of the Smithsonian Astrophysical and Harvard College Observatories.[20]

Capture theorists maintained that geochemical constraints ruled out fission, accretion from leftover Earth material in Earth orbit and twin planet development. What was left was capture. If the Moon formed separately, it had to be captured.

However, the capture hypothesis did not jibe with dynamic constraints. A body of the size and density of the Moon passing close enough to the Earth to be captured would have to come within the Roche limit, the closest it could come to the Earth without breaking up from the disruptive effect of the tides raised on it by the Earth's gravity.

The Roche limit has been calculated at 2.88 Earth radii (11,401 miles) from the center of the Earth. It is named for Edouard Roche, a nineteenth-century French mathematician.

Both of the new theories meet the capture constraint of disintegration. They simply allow their incoming body to break up. Then they reassemble it in rather ingenious fashion in the Earth's orbit.

In his theory of disintegrative capture, Smith said that the entire proto-Moon does not break up at 2.88 radii. Assuming it is already differentiated with a liquid iron core and a liquid silicate mantle, the silicates might peel off first at 3 radii whereas the denser iron core would remain intact as close as 2.24 radii, he calculated.

The iron and the silicates would be separated in this way. When the core finally broke at 2.24 radii, there would be two masses of fragments in turbulent commotion—the more massive ductile iron and the brittle silicates. Collisions between the iron and the silicate fragments would result, he said, in "preferential capture" of the more massive iron pieces. Some of these would be slowed by collision into decaying orbits that would plunge them into the Earth, but some would remain in orbit at 2.24 radii, providing the nucleus for the accretion of a baby Moon. The silicate pieces would be bounced by collisions into higher orbits. Most of the volatiles would boil off from accreting hot clouds of debris.

In this omnibus hypothesis, Smith not only gets rid of most of the iron and the volatiles, but sets the stage for the basin-forming bombardment 4 billion years ago.

His scenario calls for the reassembly of core fragments at 2.24 radii and rapid growth by the accretion of a silicate mantle as the baby Moon spirals outward to 3 radii and beyond as a result of tidal interaction with the Earth.

Still spinning, the growing Moon generates a sizable magnetic field from the dynamo effect of its rotation around a molten iron sulfide core.

As the tidal effect that causes the New Moon to recede also locks its rotation to its orbital period, the dynamo effect dies away and the magnetic field vanishes as the last of the mare basalts crystallize in the basins. After that, there is only a memory of the dynamo and of a spinning Moon in the remanent magnetism of the rocks.

Meanwhile, the clumps of silicate debris that had been bounced into the more distant orbits accrete into Moonlets orbiting the Earth at several distances. The receding Moon recaptures these pieces as it passes through their orbits. By this process, some of the volatiles and feldspars lost during disintegration are replenished, and the circular basins are formed. In its first 600 to 700 million years, the Moon has been born, shattered, and resurrected.

The modified capture theory of Wood and Mitler is less ambitious. They assume a proto-Moon that is not spinning, in contrast with the spinning Chicago Moon. It passes inside the Roche limit at parabolic velocity, which means it will keep right on going unless something happens. Something does. The proto-Moon breaks up under the tidal stress of its close approach.

At the moment of disintegration, pieces from the side nearest the Earth have less than parabolic velocity because they are closer to the Earth's center of mass than either the center or the far side of the proto-Moon. As soon as the near-side pieces are sloughed off the main body, they fall into elliptical orbits around the Earth whereas the core and the far-side fragments, which are more distant from the center of the Earth, keep traveling at escape velocity. In this way, Wood and Mitler get rid of the iron and siderophiles, or most of them.

Like the Chicago model, the Harvard-Smithsonian Moon accounts for the peculiar bulk chemical composition of the present Moon which, as Wood and Mitler pointed out, "has always been an embarrassment to those who would form it by capture or in orbit around the Earth." The embarrassment stems from the belief that the condensation of metallic iron and of magnesium silicate takes place in the same temperature interval. It is difficult to account for one without the other. "It is very awkward for the Moon to have to shun metallic particles while it is accreting magnesium silicates for its mantle," was the way the theorists put it.

Both theories have yet to show that a body coming within the Roche limit would break up on a single pass. Some experts believe that a body must actually be captured and held in an orbit before tidal effects can break it up.

Unstable, controversial, and by no means unanimous, the New Moon that rose at Houston in March 1974 is at least the most sophisticated model to come out of the Apollo program. It is incomplete because Apollo is incomplete—and also because the analysis of lunar samples, although it fills hundreds of pages, has barely started.

Of the 840 pounds of rocks and soil returned from the Moon by Apollo missions, only 5 percent or 42 pounds had been distributed to investigators in the United States at the time of the Fifth Lunar Science Conference. Only about half of that had been analyzed by that time.

Decades of lunar sample analysis lie ahead, along with further reduction of the instrument data. Thousands of photographs remain to be examined as of the spring of 1974.

As this work progresses in the United States, Canada, Latin America, Western Europe, Russia, and Japan, the rise of new models of the Moon are inevitable.

The legacy of Apollo is incalculable. The voyages to the Moon may have been the beginning of a great age or the end of one. This is a decision we ourselves must make.

Some people have made it. Co-chairing a conference session with fellow astronaut John Young, Joseph Allen remarked: "This program sounds like a planning session for the exploration of the Moon. We'll do it right next time."

Notes

1. NASA News Conference, Washington, D.C., October 27, 1972.
2. Ibid.
3. Ibid.

4. Mark Bloom in *The New York Times*, July 16, 1972.

5. NASA News Conference, Washington, D.C., October 27, 1972.

6. Ibid.

7. NASA News Conference, Cocoa Beach, Florida, December 5, 1972.

8. Report of Apollo Field Geology Investigation Team, U.S. Geological Survey, Houston, December 13, 1972.

9. Apollo 17 Preliminary Examination Team, "Apollo 17 Lunar Samples: Chemical and Petrographic Description," *Science*, Vol. 180, November 6, 1973.

10. Ganapathy, R., and Anders, E., "Bulk Composition of the Moon and Earth Estimated from Meteorites," Fifth Lunar Science Conference, Houston, 1974.

11. Grossman, L., and Clayton, R. N., "Oxygen Isotopic Compositions of Lunar Soils and Allended Inclusions and the Origin of the Moon," Fifth Lunar Science Conference, Houston, 1974.

12. Taylor, S. R., and Jakes, P., "Chemical Zoning in the Moon," Fifth Lunar Science Conference, Houston, 1974.

13. Kaula, W. M., Schubert, G., Lingenfelter, R. E., Sjogren, W. L., and Wollenhaupt, W. R., "Apollo Laser Altimetry and Inferences as to Lunar Structures," Fifth Lunar Science Conference, Houston, 1974.

14. Wasserburg, G. J., et al., "Dunite from the Lunar Highlands," Fifth Lunar Science Conference, Houston, 1974.

15. Tera, F., Papanastassious, C. J., Wasserburg, G. J., "The Lunar Time Scale and a Summary of Isotopic Evidence for a Terminal Lunar Cataclysm." Fifth Lunar Science Conference, Houston, 1974.

16. Schaeffer, O., and Husain, L., "Chronology of Lunar Basin Formation and Ages of Lunar Anorthosite Rocks," Fifth Lunar Science Conference, Houston, 1974.

17. Head, J. W., "Lunar Dark Mantle Deposits." Fifth Lunar Science Conference, Houston, 1974.

18. Morgan, J. W., Ganapathy, R., Higuchi, H., and Anders, E., "Lunar Basins. Tentative Characterization of Projectiles," Fifth Lunar Science Conference, Houston, 1974.

19. Smith, J. V., "Origin of the Moon by Disintegrative Capture," Fifth Lunar Science Conference, Houston, 1974.

20. Wood., J. A., and Mitler, H. E., "Origin of the Moon by Modified Capture," Fifth Lunar Science Conference, Houston, 1974.

Index

Accidents, on Apollo flights, 159–165, 250–252, 258–259

Accretion theory, 83, 242–243, 297–298

Agena target rocket, 63, 90–91, 249

Aldrin, Edwin E. (Buzz), Jr., 5, 7, 63*ff*, 77, 79, 91, 93, 104, 107, 155, 176, 189, 230

Allen, Joseph, 212, 213, 216, 217, 300

Alpha-particle scattering analyzer, 50–53, 54, 74, 240

Alphonsus, 210, 296

Anders, Edward, 138–139, 140–142, 242–243, 291

Anorthosite, 176–177, 230, 246, 269

Antares. *See* Apollo 14

Apennine Mountain front, 210, 228, 248

Apollo 1, fire on, 160, 163

Apollo 9, 206

Apollo 10, 6*n*, 249

Apollo 11, 63–71, 249; goals of, 43–44, 62, 91–92; Moon rocks from, 74–81; reconnaissance flights prior to, 46–64; and theories of lunar origins, 81–88

Apollo 12, 90–121; flight plan for, 79, 108–110; scientific findings of, 123–147; struck by lightning, 95–96

Apollo 13, 103, 118, 124, 125, 156–168; blowout of liquid oxygen tank on, 159–166; crew of, 155–156; and decline of space program, 168–174

Apollo 13 Review Board, 164–165, 170

Apollo 14, 167, 182–199; and biomedical studies, 203–204; Pre-Mission Science Briefing, 179–182; purposes of, 175–177, 180–182; six-iron experiment, 198; and wheeled vehicle MET, 192–193, 195

Apollo 15, 211–225; crew of, 205; and Lunar Roving Vehicle, 207; scientific findings of, 227–243; and stamp scandal, 273–274

Apollo 16, 248–269

Apollo 17, 44, 276–290; flight plan for, 275–276

Apollo 20, cancellation of, 149

Apollo Project: design of, 18; expectations for, 14–18; exploration priorities for, 39–40; knowledge about Moon after, 179–182; models of Moon prior to, 19–40; objectives of, 37–39; public criticism of, 149–151; reasons for discontinuing of, 6; results of, *ix–x*, 290–300; selection of crews for, 43–44; significance of, 6–7, 173–174, 300; stages of, 245. *See also* Apollo 1, Apollo 9, Apollo 11, *etc.*

Aquarius. *See* Apollo 13

Aristarchus Hills, 55, 209, 296

Armstrong, Neil A., 5, 7, 63, 64, 65–71, 77, 79, 80, 91, 93, 104, 107, 114–115, 176, 189, 230

Arnold, James R., 236–237

Asteroids, 24, 31

Astronauts: and biomedical experiments, 201–204; compared with explorers, 94–95; and decontamination procedure, 71. *See also under individual names of*

Atmosphere composition detector, 39, 281

Bacteria, on Moon, 110

Baldwin, Ralph B., 22–24, 97, 297

Basalt, in lunar soil, 50–53, 55, 60–61, 79, 228

Lunar surface (*cont.*)
ings on, 74–79, 245–247; formation
of, 19; glassy patches on, 79–81; and
lunar bell effect, 124; in photographs,
47–57, 224. *See also* Lunar soil
Lunar volcanism. *See* Volcanism
Luniks, 25, 100
Luny Rock I, 134, 238
Lyell, Charles, 9–10, 20, 85, 232

Magnetic field of Moon, 17, 78–79, 87–
88, 127–130, 134, 138–140, 233, 234,
268, 299
Magnetometers, 92, 125, 127–130, 193,
233, 247
Mare: Crisium, 43, 72, 231, 246; Fe-
cunditatis, 246; Humboldtianum, 59;
Imbrium, 17, 25, 30, 60, 143, 187;
Nubium, 48, 209; Orientale, 59;
Smythii, 59; Serenitatis, 25, 30, 224,
225, 276; Tranquillitatis, 17, 27, 30,
50, 66, 101, 114, 129, 132, 139, 144–
145, 177, 237, 238, 293; Vaporum,
296
Maria region, 19, 25–26, 30, 49, 50–53,
59, 77, 236
Mariner 9, 58, 84
Mars, 16, 37, 58; exploration of, 57, 92,
142, 204, 248; formation of, 46, 84–
85
Mascons, discovery of, 59–62
Mattingly, Thomas K., II, 155, 249ff
Mercury, 16, 85
Mercury programs, 46, 90, 104
Meteoric theory of Moon craters, 20–
26, 28–29
Meteorites, 31, 77, 182. *See also* Me-
teoric theory of Moon craters
Minerals on Moon, 76, 131, 133–134
Mitchell, Edgar, 167, 183, 203–204, 237,
258
Mitler, H. E., 298–300
Modularized equipment transporter
(MET), 192–193, 195
Moho on Moon, 181
Mohorovičić discontinuity. *See* Moho
Moment of inertia of Moon, 138, 292–
293
Month, lengthening of, 34–35
Moon: age of, 45–46, 77–78, 134, 143–
144, 229, 240–243, 271–272, 294–
296; chemical and mineral structure
of, 16; distance between earth and,
69; existence of water on, 27–28;
morning on, 100–101; melting of, 85–
87; planetesimal bombardment of,
241–243; and solar storms, 102–103;
temperature of, 59–62, 71, 78–79,
85–87, 93–94, 131–132; thermal cycle
of, 29–30; volatile zones of, 297–298.

See also Moon models *and entries
under* Lunar
Moon models: and Apollo findings, 81–
88, 227–243, 290–300; of Baldwin,
22–24; conflicting, 44–46; of Darwin,
33–36; of Kuiper, 29–31; of Moore,
24–26; of Opik, 31–33; petrologic,
85–87; of Taylor and Jakes, 292–294;
of Wood and Mitler, 299–300
Moon origins. *See* Lunar origins
Moonquakes, 72–74, 93, 125–126, 182,
209, 228, 231
Moon rocks: from Apollo 11, 64, 74–
79; from Apollo 12, 113, 118, 130–
135; from Apollo 16, 257; from
Apollo 17, 290; chemical analysis of,
85–86; difference between age of fines
and breccia and, 140–145; distribution
of, 237–230; and *E* component, 142,
143–145; from Fra Mauro, 176–177;
future analysis of, 300; and hetero-
geneity of Moon, 235–237; of North
Bay crater, 267–268; and origin of
Moon, *see* Lunar origins; radiometric
dating of, 11–12; from South Ray
Crater, 216–262; types of, 228
Moon seas, 25, 52
Moon soil. *See* Lunar soil
Moon suits, 16–17, 38, 212
Moore, Patrick, 24–26, 95
Mt. Hadley Delta, 217–220, 221–223

NASA, 6, 13–14, 54, 57; effects of bud-
get cuts on, 205; goals of, 43–44;
scandal in, 273–274. *See also* Apollo
National Aeronautics and Space Admin-
istration. *See* NASA
Neptune, 16, 37
News media, 104, 170, 172, 179–182.
See also Television
Nixon, Richard M., 66, 151, 170–171,
221

Ocean of Storms, 107, 115, 128, 130,
144, 208
Oceanus Procellarum, 22, 25, 49, 55,
58, 95, 99, 100, 108–113, 128, 131,
209, 230, 231, 235, 237, 238, 272, 296
Odishaw, Hugh, 14, 17
Odyssey. *See* Apollo 13
O'Keefe, John, 35–36, 37, 82, 83, 135–
136
Opik, Ernst J., 31–33, 86, 137
Orange glass, 296–298
Orange juice dispenser, 252–253, 256,
259, 264
Orbiter project, 46–47, 51
Orion. *See* Apollo 16
Oxygen, 54; in lunar soil, 51; and walk-
ing radius on Moon, 38–39
Oxygen tanks, 159–166, 175